符号中国 SIGNS OF CHINA

老茶具

ANCIENT TEA SETS

"符号中国"编写组 ◎ 编著

中央民族大学出版社
China Minzu University Press

图书在版编目(CIP)数据

老茶具：汉文、英文／"符号中国"编写组编著．— 北京：
中央民族大学出版社，2024.8
（符号中国）
ISBN 978-7-5660-2339-1

Ⅰ.①老… Ⅱ.①符… Ⅲ.①茶具—介绍—中国—汉、英 Ⅳ.①TS972.23

中国国家版本馆CIP数据核字（2024）第017030号

符号中国：老茶具 ANCIENT TEA SETS

编　　著	"符号中国"编写组
策划编辑	沙　平
责任编辑	周雅丽
英文指导	李瑞清
英文编辑	邱　械
美术编辑	曹　娜　郑亚超　洪　涛
出版发行	中央民族大学出版社
	北京市海淀区中关村南大街27号　邮编：100081
	电话：（010）68472815（发行部）　传真：（010）68933757（发行部）
	（010）68932218（总编室）　　　　　（010）68932447（办公室）
经销者	全国各地新华书店
印刷厂	北京兴星伟业印刷有限公司
开　　本	787 mm×1092 mm 1/16　印张：10.75
字　　数	154千字
版　　次	2024年8月第1版　2024年8月第1次印刷
书　　号	ISBN 978-7-5660-2339-1
定　　价	58.00元

版权所有　侵权必究

"符号中国"丛书编委会

唐兰东　巴哈提　杨国华　孟靖朝　赵秀琴

本册编写者

文　铮

前言 Preface

中国是茶的故乡，茶文化经过几千年的发展，已经成为了中国人血脉中的特殊印记。而作为茶文化的重要载体，茶具与茶文化共生、共存，共同发展，同样也具备了独特而深厚的文化内涵。爱茶的人，必然也爱那些古朴而美丽的老茶具。透过老茶具，可以体会到古人饮茶的生活情趣，今人与古人嗜茶的心也在这一刻相通、共鸣。

China is the home of tea. The tea culture has become a special mark on Chinese people in the past thousands years. Tea sets are an important carrier of tea culture. Appearing together with tea culture, existing and developing along, teaware also has special and profound culture connotation. Those who love tea must also be interested in the simple, unsophisticated and beautiful old teaware. We could feel the lifestyle of those ancient people who enjoyed drinking tea. People at present can also have the same feelings for tea as ancient people.

When did the first tea set come out? What are the differences between teawares in ancient times and today? Why did people in the

中国第一件专用的茶具是什么时候产生的？古人使用的茶具和今天又有什么不同？宋代人为什么偏爱黑釉盏？紫砂壶为什么在明清时期风靡一时？为了令读者对中国老茶具有一个全面的了解，本书详细介绍了老茶具的发展历史，以及各种材质老茶具的基本知识。希望读者能够在了解中国老茶具的同时，体味纯正的中国味道，领略博大的中国文化。

Song Dynasty prefer black-glazed bowl? Why were purple clay teapots so popular in the Ming and Qing dynasties? In order to present readers with an overall idea of old Chinese teawares, this book introduces the history of traditional teawares and the basic knowledge of their different materials in detail. We hope as readers know more about traditional Chinese teawares, they can also understand true Chinese style and appreciate this culture.

目录 Contents

茶俗与茶具的发展
The Development of Tea Customs and Teawares 001

唐代以前的茶俗与茶具
Tea Customs and Teawares before the Tang Dynasty 002

唐代的茶俗与茶具
Tea Customs and Teawares in the Tang Dynasty 008

宋代的茶俗与茶具
Tea Customs and Teawares in the Song Dynasty 027

元代的茶俗与茶具
Tea Customs and Teawares in the Yuan Dynasty 044

明清时期的茶俗与茶具
Tea Customs and Teawares in the Ming and Qing Dynasties 055

各种材质的老茶具
Ancient Teawares of Different Materials 075

瓷茶具
Porcelain Teawares 076

紫砂茶具
Purple Clay Teasets 100

金银茶具
Gold and Silver Teawares 117

琉璃茶具
Glass Teawares 122

漆茶具
Lacquer Teawares 124

锡茶具
Tin Teawares 127

珐琅茶具
Enamel Teawares 130

玉石茶具
Jade Teawares 132

竹木茶具
Bamboo-wood Teawares 138

果壳茶具
Nutshell Teasets 143

老茶具的收藏与保养
Storage and Maintenance of Ancient Teawares 145

瓷茶具的收藏与保养
Storage and Maintenance of Porcelain Teawares 146

紫砂茶具的收藏与保养
Storage and Maintenance of Purple Clay Teasets 148

金属茶具的收藏与保养
Storage and Maintenance of Metal Teawares 153

漆茶具的收藏与保养
Storage and Maintenance of Lacquer Teawares 155

竹木茶具的收藏与保养
Storage and Maintenance of Bamboo-wood Teawares 157

- 紫砂东坡提梁壶（清）
Purple Clay *Dongpo* Tea Pot with Loop Handle (Qing Dynasty, 1616-1911)

茶俗与茶具的发展
The Development of Tea Customs and Teawares

　　茶具是中国人使用的历史最悠久的日用器具之一。几千年来，随着制茶技术、饮茶习俗和茶文化的发展，茶具在材质、器形、种类上也发生着相应的变化。

Teaware is one of the daily utensils that Chinese people have used for a long time. With the development of tea making techniques, tea customs and tea culture, teaware has changed correspondingly in materials, shapes and types in the past thousands years.

> 唐代以前的茶俗与茶具

众所周知，中国是茶的故乡。茶树原本是一种野生的植物，最初人类并不知道它的功能与用途。新石器时代晚期，中国人最早发现了茶叶并开始了饮茶的历史。

- 神农氏画像
Portrait of Shennong

> Tea Customs and Teawares before the Tang Dynasty

China is well-known as the hometown of tea. The tea tree was originally a wild plant and people did not know its functions and uses at the beginning. In the late stage of the Neolithic Period, Chinese people were the first to discover tea and thus began the history of tea drinking.

There is a legend about the origin of the tea custom. It is said that in the primitive society about 5000 years ago, there was Shennong, a leader of a Central Plains tribe. He invented ploughs and plowshares for farming and taught people to grow food. He was the first to discover herbal medicine and help people to heal wounds. Naturally, Shennong was also the first person to discover tea. He found out that this plant was not only delicately scented, but also had the function of thirsty quenching, mind refreshing, urine inducing

关于饮茶习俗的起源，还有一个传说。据说在距今5000多年前的原始社会，中原有位部落首领，被称为"神农氏"，他最早发明了耕地用的耒耜，教会人们种植粮食；

and detoxicating … This is just a legend, but it reflects the fact that Chinese people discovered and used tea thousands of years ago.

People began to settle down and make potteries in the era of Shennong. At that time, they only realized the medicinal value of tea, so they often simply chewed them. Therefore, there was no teaware for special use then. Ceramic pots and

- **云南镇沅千家寨茶树王**
  云南镇沅县千家寨遗址北部约2千米处的原始密林中，有上万亩的散生野茶树林，其中一株高18.5米、胸围2.82米的野茶茶树，据考证已有2700年的树龄，是迄今发现最古老的野茶树，堪称稀世之宝。

  The King of Tea Trees in Qianjia Village, Zhenyuan County, Yunnan Province
  In the primitive forest two kilometers from the Qianjia Village Site, Zhenyuan County, Yunnan Province, there are thousands of scattered and wild tea trees. One of them is a wild tea tree with the height of 18.5 meters and the perimeter of 2.82 meters. 2700-year-old and an absolutely rare treasure, it is the oldest wild tea tree discovered yet.

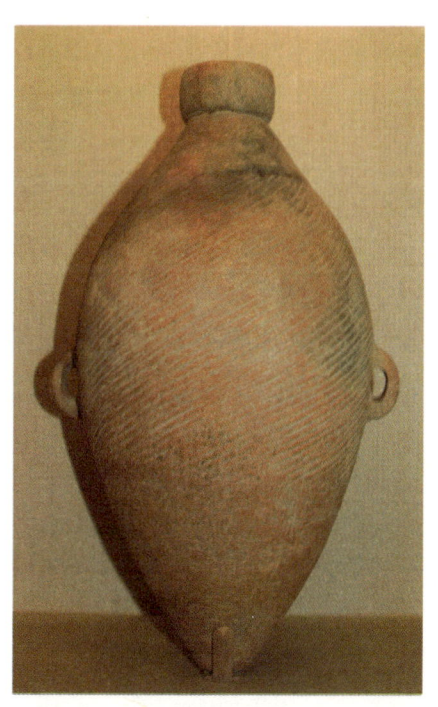

仰韶文化小口尖底陶瓶（新石器时代晚期）
Small Mouth, Pointed Bottomed Ceramic Bottle from Yang Shao Culture (Late Period of Neolithic Period)

他最早发现了草药，帮人治病疗伤。当然，神农氏也是最早发现茶叶的人。他发现这种植物不仅味道清香，还能解渴生津、提神醒脑，并有利尿解毒的作用……这虽然只是个传说，但反映出中国人发现与利用茶叶的历史至今已有数千年。

　　神农氏的时代，人们已经开始制造陶器，并过上了相对稳定的定居生活。当时人们对茶的认识仅停留在药用价值的阶段，通常是简单地咀嚼食用，因此还谈不上使用专用的茶具。新石器时代的陶罐、陶钵，可以看成是茶具的源头，因为古人除咀嚼茶叶这一种方法外，也有可能把茶叶和其他蔬菜一起混煮或混用。直到今天，中国云南省西双版纳的基诺族的餐桌上还有凉拌茶这道传统菜肴。

● 陶碗（战国）
Ceramic Bowl (Warring States Period, 475 B.C.-221 B.C.)

bowls in Neolithic Period can be taken as the origin of teaware, because besides chewing tea, ancient people might cook and use it with other vegetables. Until today, the Jino ethnic group in Xishuangbanna, Yunnan Province, also has this traditional cold dish.

　　In the Shang and Zhou dynasties (1600 B.C.-256 B.C.), historical books did not record much about tea. Only a few mentioned that people in Ba-Shu area (now Sichuan Province) grew tea trees and offered tea as tribute to the Emperor of Zhou. In addition, Ba-Shu area is also the origin to take tea as a drink. In the late time of the Warring States Period (about 221 B.C.), the Qin State defeated small states in Ba-Shu area, and then tea popularized through the country.

　　From the Han Dynasty (206 B.C.-220 A.D.) to the period of the Three Kingdoms and the Southern and Northern dynasties (220-589), there were more records about the use of tea. People had very different ways of drinking tea at that time. Fresh tea leaves would be smashed, dried and mixed with rice cream to solidify. Then they were reserved as tea cakes. When people wanted to drink tea, they should bake the tea cake till it became red, smash it to powder and

商周时期，史籍中关于茶叶的记载并不多，只是零星地提到巴蜀一带（现四川省境内）种植茶树，巴人曾把茶叶作为贡品献给周天子。将茶作为饮料，也是始于巴蜀地区。战国末期（前221年左右），秦灭巴蜀以后，茶才在全国逐渐普及开来。

从汉代（前206—公元220）到三国两晋南北朝，关于用茶的记载开始多起来。那时人们的饮茶方式同现在有很大的区别。鲜茶叶采下来后，要捣碎、烘干，并且加入米膏之类的食物使之凝固，制成茶饼备用。每次饮用前，要先把茶饼放在火上烤成赤红色，然后捣成粉末，放入容器中，再注入开水冲饮，有时还要加入葱、姜之类的调

then put it in containers. People drank tea by mixing them with boiling water and sometimes put scallion and ginger as seasonings. They had special teawares to make tea cakes, bake tea, smash tea or drink it. The most important teaware was the tea bowl, and there was the tea bowl base at that time. It is worth mentioning that there were a number of ceramic and porcelain teawares in this period, especially celadon teaware from *Yue* Kiln. This is because the prosperous development of *Yue* Kiln in the southern part of China. Celadon from the South plays an important role in the history of

- 越窑青瓷鸽形杯（西晋）
Celadon Tea Cup with Dove Pattern from *Yue* Kiln (Western Jin Dynasty, 265-316)

- 青瓷盏托（东晋）
东晋时越窑生产的青瓷盏托，釉色滋润如玉，釉面经过近2000年时间的浸浮而微微有细小开片。当时的盏和托盘是粘连在一起的，如此设计的优点是喝茶时盏不易滑脱。

Celadon Bowl and Base (Eastern Jin Dynasty, 317-420)
The celadon bowl and base from *Yue* Kiln in the Eastern Jin Dynasty (317-420) were glazed like jade. After nearly 2000 years, the glazed surface has minor crackle. The bowl and base were stuck together at that time. This design keeps the bowl from falling from the base when tea is served.

味品。而制茶饼、烤茶、碾茶以至饮用，都有了相应的茶具，其中最主要的茶具就是茶碗，并且已出现了茶盏托。值得一提的是，陶瓷茶具，特别是越窑青瓷茶具在这一时期大量出现。分析原因，是同这一时期南方越窑的繁荣发展分不开的，南方青瓷在中国古陶瓷的发展史上具有举足轻重的作用。瓷器的耐高温、光润的釉面以及易清洗的特点，相对陶器来说，其优势明显凸现，所以得到越来越多的人接受。不仅贵族接受它，百姓也是把瓷器作为日常生活不可或缺的器皿，所以这个阶段以陶瓷制作的茶碗、茶盏托才会大量出现。

traditional Chinese porcelains. Porcelain is heat-resistant, has a smooth glaze and is easy to wash. Compared with ceramic utensils, porcelains have more advantages and are accepted by more people. Not only the noble people liked them, but also common people took them as necessary daily utensils. Therefore, there appeared a large number of ceramic and porcelain tea bowls and bases.

开片

　　开片，本为瓷器釉面的一种自然开裂现象，原是瓷器烧制过程中出现的失误和缺陷。但人们发现釉面布满裂纹有一种特殊的美感，并逐渐掌握了釉面开裂的规律，故意制出了开片釉（裂纹釉），使其成为了瓷器的一种特殊装饰。宋代的汝窑、官窑、哥窑都有这种开片产品，以哥窑产品最为著名。开片釉按颜色分有鳝血、金丝铁线、浅黄鱼子纹；按形状分有网形纹、梅花纹、细碎纹等。

Crackle

Crackle, a natural phenomenon of glazed porcelain surface cracking used to be formed by mistake and was a defect in the process of firing porcelains. However, people noticed that the glazed surface had a special kind of beauty with cracks, so they began to learn to make Crackle Glaze on purpose as a unique decoration of porcelains. In the Song Dynasty (960-1279), *Ru* Kiln, Official Kiln and *Ge* Kiln all had these Crackle products, among which *Ge* Kiln was the most famous one. Crackle glaze is categorized by color as eel blood, gold and iron wire, and light yellow fish egg lines, while by shape there are the web pattern, the plum blossom, and broken bits patterns.

- **仿哥釉碗（清）**

　　宋代以后，哥窑瓷器在历代都有仿烧，被称为"仿哥釉"，但在艺术效果上无一能与宋代哥窑相媲美。

Imitated *Ge* Kiln Glazed Bowl (Qing Dynasty, 1616-1911)

After the Song Dynasty, every dynasty imitated porcelains from *Ge* Kiln. These imitations are called "Imitated *Ge* Kiln". However, none of them can match the artistic effect of those from *Ge* Kiln in the Song Dynasty (960-1279).

> 唐代的茶俗与茶具

唐代是一个相当开放的历史时期，经过开国几代帝王的励精图治，到了开元、天宝年间，大唐的物质财富达到鼎盛期。人们生活富裕，对精神生活的需求日益加剧，此时的饮茶已不仅仅停留在粗放式解渴、药饮的层面，而是追求艺术化的过程，也就是后人所说的"品饮阶段"。

唐人的煮茶习俗

唐代的茶叶种植面积大增，产量也大幅度提高。由于南北气候悬殊，在不同的地理、气候环境下，各地生产的茶叶质量也不尽相同，通过时人的评比，产生了不少名茶。由于茶叶产量以及消费量的上升，朝廷也开始把茶作为征税的重

> Tea Customs and Teawares in the Tang Dynasty

The Tang Dynasty (618-907) is a relatively open period in ancient China. After the hard work of several dynasty founders, the Tang Dynasty reached the height of great material wealth in Kaiyuan and Tianbao periods. People lived a prosperous life, so they had an increasing pursuit for spiritual life. In this period, drinking tea did not only mean roughly satisfying thirst or for medication. It became a process of pursuing art, which was also named by later generations as the "tasting period".

The Custom of Tea Making in the Tang Dynasty

People in the Tang Dynasty grew more tea than before, and there saw a great

要对象，茶税，成为唐代的一项重要财政收入。同时朝廷在顾渚（今江西宜兴）设立贡茶院，专门派人监督加工贡茶。最好的茶叶都集中由皇室控制，如果当时官员政绩颇佳或是做出了特殊贡献，以及外邦来朝，都会受到大唐天子的赐茶礼遇。

相对而言，唐人饮茶的程序比我们现代人复杂得多，从另一方面说，也艺术得多。这主要是由茶叶的加工方法不同决定的。唐代的茶

- 白釉茶炉及茶釜（唐）
因唐代的饮茶方式以烹煮为主，所以茶釜是当时的一种重要茶具。
White-glazed Tea Stove and Boiler (Tang Dynasty, 618-907)
Since people in the Tang Dynasty mainly boiled or fried tea, tea boiler was an important teaware at that time.

increase in tea output. Due to the big difference of weather in northern and southern parts of China, the qualities of tea in different places varied a lot under different geographical and climatic conditions. High quality teas stood out after people's comparison and assessment. The imperial court began to take tea as an important object to collect tax as its output and consumption rose. Tea tax was a main source of state revenue. At the same time, the imperial court opened Tribute Tea Station in Guzhu (now Yixing, Jiangsu Province) and dispatched officials to monitor the processing of tribute tea. The best teas were all controlled by the royal family. If officials performed outstandingly in work, or when foreign guests made state visits, the emperor would award them with tea.

People in the Tang Dynasty had a relatively sophisticated process of drinking tea. In other words, their way of drinking tea had more sense of art. This is mainly decided by different ways of making tea. Cake tea was the main form then. People paid particular attention to "boiling" or "frying" their tea. Lu Yu, the sage of tea, recorded 28 teawares for different uses in *The Classic of Tea*.

叶以饼茶为主,茶叶品饮方式讲究"煮茶"或"煎茶"。"茶圣"陆羽在《茶经》中曾经记载了28种用途各异的茶具,而在普通人家里,准备的茶具可能相对简单些,但其中几件则是必须的,那就是茶炉、茶釜、茶碾(或茶臼、茶磨)、茶碗。煮茶的程序是,先把饼茶放在火上炙烤片刻,后放入茶臼或茶碾中碾成茶末,放入茶罗筛选,将筛

In ordinary families, people might just have less tewares. However, some of the teawares are essential: tea stove, tea boiler, tea roller (or tea mortar, tea mill) and tea bowl. To make tea, first people will bake the tea cake for a while, then put it into tea mortar or mill, grind it into tea powder, filter the powder in a sieve and keep it in the tea box. On the other hand, people also need to prepare the stove, put an appropriate amount of water in the tea boiler. When the water is boiling for the first time (as the size of bubbles looks like crab eyes), add some salt according to the amount of water.

- "陆羽品茶"彩塑

陆羽(733—约804),字鸿渐,唐朝复州竟陵(今湖北天门)人。他是个孤儿,在佛寺中长大,却不愿诵经念佛,只酷爱茶艺。成年后,陆羽经过多年的游历,考察各地种茶、采茶、煮茶、饮茶的习俗,倾心研究饮茶之道,终于写成了《茶经》。

Color Modeling of "Lu Yu Tasting Tea"

Lu Yu (733-approx. 804), also known as Hongjian, was from Jingling, Fuzhou (now Tianmen, Hubei Province). He was an orphan and grew up in a temple. There he did not chant much sutra but only focused on making tea. When he became an adult, he traveled a lot to study the customs of growing, collecting, making and drinking tea in various places. He made all his efforts to study tea and finally wrote the book *The Classic of Tea*.

• 元 赵原《陆羽烹茶图》
这幅图中山水清远,茅屋数座,屋内倚坐在榻上的人即陆羽,前有一童子正在焙炉烹茶。
Lu Yu Making Tea by Zhao Yuan (Yuan Dynasty, 1206-1368)
At the far back of the picture are mountains and rivers. There are several thatched cottages. Lu Yu is the one who leaned against the coach in the cottage. In the front of him, a boy is making tea on the stove.

出的茶末放在茶盒中备用。另外还要准备好风炉,茶釜中放入适量的水,煮水至初沸(观察釜中水泡如蟹眼)时,按照水量的多少放入适量的盐;到第二沸(釜中水泡如鱼眼)时,用勺子舀出一勺水储放在熟盂中备用;釜中投放适量的茶末,等到第三沸(釜中水如波浪翻腾)时,把刚舀出备用的水重倒入茶釜,使水不再沸腾。这时茶就已煮好了,准备好茶碗,把煮好的茶用勺子注入茶碗。

When the water is boiling for the second time (the size of bubbles looks like fish eyes), scoop out a tea spoon of water and keep it in jar. Put an appropriate amount of tea powder in the boiler. When the water is boiling for the third time (now the water is rolling over), put the reserved water back into the boiler, in order to stop the boiling. Tea is ready at this time. Then prepare tea bowls and fill them with tea spoon.

唐代生产瓷茶具的主要窑场
Main Teaware Kilns in the Tang Dynasty

越窑

越窑青瓷起源于魏晋南北朝时期,主要产地是在浙江省的宁绍地区(唐代称"越州"),这里是中国主要的青瓷发源地。唐代的文人雅士喜欢饮茶,而越窑青瓷温润如玉的釉质、青绿略带闪黄的色彩能完美地烘托出茶汤的绿色。因此越窑青瓷受到了文人雅士的喜爱。

Yue Kiln

Celadon from *Yue* Kiln started in the Wei Dynasty (220-265), the Jin Dynasty (265-420) and the Southern and Northern dynasties (420-589). The main place of production was Shaoxing and Ningbo area in Zhejiang Province (called Yuezhou in the Tang Dynasty). This is also the major place of origin of celadon. Refined scholars in the Tang Dynasty loved drinking tea. Celadon from *Yue* Kiln has the glaze like jade and color combined green with shining yellow, which perfectly sets off the green color of the tea infusion by contrast. Therefore, celadon from *Yue* Kiln was loved by many refined scholars.

● 越窑青瓷横把壶(唐)
Yue Kiln Celadon Pot with Horizontal Handle (Tang Dynasty, 618-907).

邢窑

邢窑也是唐代重要制瓷窑口之一,位于河北省内丘,唐时属邢州,所以又称为"邢窑"。邢窑以烧制白瓷著称。碗是唐代白瓷中最流行、出现最多的器形,随着饮茶之风的盛行,邢窑白瓷茶碗成为常见日用瓷器,器形一般小而浅,略似斗笠,敞口浅腹,比较厚重,口沿有凸起的厚卷唇。

Xing Kiln

Xing Kiln, located in Neiqiu, Hebei Province, was also one of the most important porcelain kilns in the Tang Dynasty (618-907). Neiqiu belonged to Xingzhou in the Tang Dynasty, so the kiln here was called "*Xing* Kiln". *Xing* Kiln was famous for its white porcelain. Bowl was the most popular and frequently seen shape

● 邢窑白瓷茶碗(唐)
White Porcelain Bowl from *Xing* Kiln (Tang Dynasty, 618-907)

in white porcelain of the Tang Dynasty. As drinking tea grew more popular, white porcelain bowl from *Xing* Kiln became household teaware at that time. The bowls are usually small ones, shallow, heavy, shaped like bamboo hats. Their rims are thick and curled.

寿州窑

寿州窑是我国唐代著名瓷窑之一，位于今安徽省淮南，因唐代时属寿州，故名"寿州窑"，始烧于隋代。唐代是寿州窑发展的繁荣期，主要烧黄釉瓷和少量黑釉瓷，产品有碗、盘、杯、盏等。唐代寿州窑瓷器的主要特征是胎色白中泛黄，釉色以黄色为主，釉面光润透明。

Shouzhou Kiln

Shouzhou Kiln was one of the most famous porcelain kilns in the Tang Dynasty, located in Huainan, Anhui Province. *Shouzhou* Kiln was so named because that place belonged to Shouzhou in the Tang Dynasty. They started to make porcelain in the Sui Dynasty (581-618), and the Tang Dynasty saw a prosperous period of *Shouzhou* Kiln. Yellow-glazed porcelain and a small amount of black-glazed porcelain were its main products, like bowls, plates, and cups. *Shouzhou* Kiln porcelains were yellow within white, yellow as the main color, with a bright and transparent surface.

• 寿州窑瓷汤瓶（唐）
Tea Pot from *Shouzhou* Kiln (Tang Dynasty, 618-907)

洪州窑

洪州窑位于江西省南昌郊县丰城市境内，从东汉开始烧制瓷器，釉色主要以青绿釉和黄褐釉为主。洪州窑青瓷的装饰以各类莲瓣纹为主流。唐代时洪州窑生产大量的民用茶具，釉色以黄褐色为主。

Hongzhou Kiln

Hongzhou Kiln is in Fengcheng town, Nanchang, Jiangxi Province. It started making porcelain since the Eastern Han Dynasty (25-220). The main color of glaze in *Hongzhou* Kiln porcelain is green and yellow. Various kinds of lotus petals were the main stream of celadon in *Hongzhou* Kiln. In the Tang Dynasty (618-907), *Hongzhou* Kiln produced a large number of teawares for civil use, with yellowish-brown as its main color of glaze.

• 洪州窑褐绿釉团花纹茶碗（唐）
Hongzhou Kiln Tea Bowl with Brown and Green Grains (Tang Dynasty, 618-907)

婺州窑

婺州窑位于今天的浙江省金华地区，唐代属婺州，故名"婺州窑"，以生产青瓷为主。唐代婺州窑生产的黑褐釉及青釉褐斑蟠龙纹瓶、多角瓶很有特色。

Wuzhou Kiln

Wuzhou Kiln is now located in Jinhua area, Zhejiang Province. It belonged to Wuzhou in the Tang Dynasty (618-907), thus was named "*Wuzhou* Kiln". Celadon was the main product of *Wuzhou* Kiln. In the Tang Dynasty, *Wuzhou* Kiln made loong-pattern bottles and bottles of many horns with black-brown glaze, or celadon with brown spot, which are very unique.

长沙窑

长沙窑是唐代南方重要的大规模青瓷窑场，位于湖南省长沙市郊铜官镇瓦渣坪。长沙窑最重要的成就，是最先把铜作为高温着色剂应用到瓷器装饰上，烧出了以铜红作为装饰的彩瓷，这是我国陶瓷史上的一项重大发明。长沙窑的产品种类较多，包括大量茶具。此外，唐人煮茶常用的茶臼、茶碾和研钵等茶具，在长沙窑遗址也时有出土。

Changsha Kiln

Changsha Kiln was an important big celadon kiln of the South in the Tang Dynasty (618-907), located in Wazhaping, Tongguan town, Changsha, Hunan Province. *Changsha* Kiln had a great achievement that they were the first to apply high temperature colorant to porcelain decoration and made faience with copper red decoration. This was a big invention in Chinese ceramic and porcelain history. *Changsha* Kiln made various kinds of products, including a great number of tewares. In addition, people also discovered many tewares like mortar and roller in the relics of *Changsha* Kiln.

● 婺州窑褐斑罐（唐）
Brown Spot Pot from *Wuzhou* Kiln (Tang Dynasty, 618-907)

● 长沙窑黄釉斑彩执壶（唐）
Yellow-glazed Colored Pot from *Changsha* Kiln (Tang Dynasty, 618-907)

唐代主要茶具类型

茶叶的消费也推动了茶具的发展。唐代茶具中最受欢迎的是瓷质茶具。自东汉后期成熟的瓷器产生后，瓷器就以其耐高温、产量大、价格低廉、易清洁等特点受到大众的欢迎，而且成为茶具的主要材质之一。除了瓷质茶具之外，金银、琉璃茶具在宫廷和贵族中也十分风行。唐代的茶具有以下几种典型器。

茶釜

茶釜是当时的一种重要茶具，因唐代的饮茶方式以"烹煮"为主，要把茶饼碾成末放入茶釜中煎煮。在越窑青瓷中，茶釜也曾大量出现，为了解唐代的煮茶方法提供了实物依据。

Main Teaware Types in the Tang Dynasty

Consumption of tea also promoted the development of teawares, among which porcelain was the most popular in the Tang Dynasty. Early in the late stage of the Eastern Han Dynasty, porcelain production was well-developed. With the features of heat-resistant, mass-producible, inexpensive and easy to clean, porcelains were very popular and became the main type of teawares. Besides porcelains, gold, silver and glass tea sets were much welcomed by royal people. Here are some classic teawares at that time.

Tea Boiler

Tea boiler is an essential part of teawares because "frying" and "boiling" were the main methods to make tea and people had to use them to boil tea powders from tea cakes. There are also a large number of tea boilers in celadon from *Yue* Kiln, which provides a material evidence for the study of tea making in the Tang Dynasty.

- 越窑青瓷釜（五代）
Celadon Tea Boiler from *Yue* Kiln (Five Dynasties, 907-960)

茶臼

茶臼是一种将茶叶磨成粉末的器皿。陆羽在《茶经》中提到的用来研磨茶叶的工具是茶碾，而茶臼出现的时间比茶碾更早。唐代一些大诗人的诗作就有不少提到过茶臼。茶臼的臼体坚致厚实，平底，外面施釉，而臼里露胎，不施釉，而且布满月牙状的小窝，坑坑洼洼，正好用来研茶。

Tea Mortar

Tea mortar is a utensil for grinding tea into powder. Lu Yu mentioned tea roller for tea grinding in *The Classic of Tea*, yet tea mortar was used even earlier. Some well-known poets in the Tang Dynasty talked about tea mortar in their poems. The body of tea mortar is solid, flat bottomed, with a glaze on the outside but not inside. Crescent-like pits spread all over inside which suit well for grinding tea.

Tea Weighing Scoop

Tea weighing scoop serves as a measuring tool, to measure the amount when tea powder is put into the boiler. Celadon tea weighing scoops from *Yue* Kiln were discovered in archaeological excavation.

- 白瓷茶臼（唐）
White Porcelain Tea Mortar (Tang Dynasty, 618-907)

茶则

量器的一种，茶末入釜时，需要用茶则来量取。用越窑青瓷制作的茶则也在考古发掘中出现。

- 越窑青瓷茶则（唐）
Tea Weighing Scoops of Celadon from *Yue* Kiln (Tang Dynasty, 618-907)

茶瓯

茶瓯是最典型的唐代茶具之一，又分为两类，一类是以玉璧底碗（圈足宽大，中心内凹，近似玉璧）为代表，另一种常见的茶碗是花口，通常作五瓣花形，腹部压印成五棱，圈足稍外撇，这种器形出现要略晚于玉璧底型。

- 越窑青釉海棠式碗（唐）
Begonia-shape Bowl of Celadon from *Yue Kiln* (Tang Dynasty, 618-907)

茶托

茶托是防茶杯烫手而设计的器形，在东晋时就有青瓷盏托出现。唐代茶托的造型较两晋南北朝时更加丰富，莲瓣形、荷叶形、海棠花形等各种款式的茶托均大量出现。

- 越窑青瓷托盏（唐）
Tea Base and Bowl of Celadon from *Yue Kiln* (Tang Dynasty, 618-907)

Tea Bowls

It is one of the most classic teawares in the Tang Dynasty. There are mainly two groups of tea bowls. One is the jade-ring bowl (wide, hollow in the centre, looking like a at jade ring). The other is featured with a flower-like mouth and normally has five petals. Its middle part has five ridges and the edge goes slight outward. This type came out later than the jade-ring-like type.

Tea Base

Tea base is designed to prevent people from being scalded by hot tea and it firstly showed up in the Eastern Jin Dynasty (317-420). Tea bases came in more shapes in the Tang Dynasty (618-907) than in the Jin Dynasty (265-420) and the Southern and Northern dynasties (420-589), with a great number of lotus-shaped and begonia-shaped tea bases.

汤瓶

到了晚唐五代之际，饮茶方式又有了新的变化，点茶开始出现。点茶需用的茶具是汤瓶，又叫"偏提"，是从盛酒用的器皿酒注子演变而来的。以汤瓶盛水在火上煮至沸腾，置茶末于碗、盏中，再将汤瓶中的沸水注入茶碗。唐人点茶时非常讲究使用汤瓶注汤的技巧。

由于无论是煎茶还是点茶，都需要将茶饼碾成末，所以除以上几种茶具之外，茶碾、茶罗、茶盒也是唐代不可或缺的茶具。

Tea Pot

In the late stage of the Tang and Five dynasties, tea drinking changed as there was a new way of making tea — tea dripping. This way of tea making required a "pot with handle at side", which was developed from wine containers. When water is boiled in the pot, put tea powder into tea bowls and pour boiling water in them. People in the Tang Dynasty gave much attention to the technique of pouring water.

As both boiling tea and tea dripping needed to grind tea into powder, the tea roller, the tea sieve and tea case were all necessary teawares in the Tang Dynasty.

• 越窑青釉执壶（唐）
Celadon Pot with Handle from *Yue* Kiln
(Tang Dynasty, 618-907)

《茶经》中记载的茶具

《茶经》成书于唐上元元年（760年），由著名茶人陆羽所著。这是一部关于茶叶生产的历史、源流、现状、生产技术，以及饮茶技艺、茶道原理的综合性论著。此书不仅是一部精辟的农学书籍，更是一本阐述茶文化的专著。《茶经》中记载的品饮方式很复杂，光列举的茶具就达28种之多，大致包括煮水工具、碾末工具和盛储工具。

Teawares Recorded in *The Classic of Tea*

The Classic of Tea written by Lu Yu, the famous tea scholar, was finished in the first year of Shangyuan Period (760) of the Tang Dynasty. It is a comprehensive book about the history, origin, status quo, production technique, customs and theories of tea. This book is not just an incisive agronomical book, but also a monograph about tea culture. The way of drinking tea described in this book is very complicated. Just in the part of teawares, this book lists 28 different types, including utensils for boiling water and grinding and storing tea.

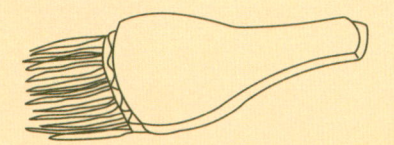

- 札，饮茶后清洗茶具的用具。
  Zha is a tea ware scouring brush.

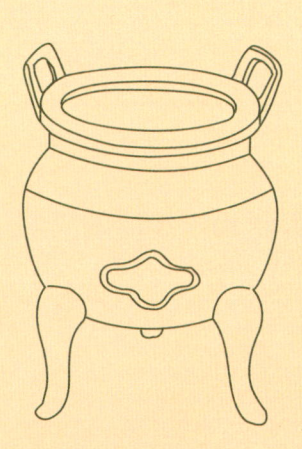

- 风炉，煎、煮茶用的小火炉，一般用铜、铁铸造。
  Stove is used for tea frying and boiling, often made of copper or iron.

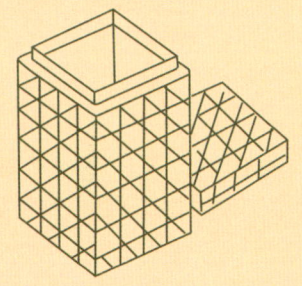

- 莒，用来盛放风炉的器具，用竹子或藤编织。
  Bamboo basket is used to store stove, made by bamboo or vine.

- 灰承，承放炉灰的器具。

  Ash tray is a utensil for ash.

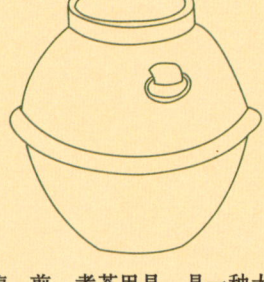

- 镬，煎、煮茶用具，是一种大口锅。将茶叶碾碎后放入其中煎煮，可用铁、银、石、瓷制成。

  Cauldron is a big mouth pot for tea frying and boiling after grinding, made of iron, silver, stone and porcelain.

- 拂末，用来扫拂茶粉的器具。

  Tea broom is used to clean the tea powder.

- 夹，夹茶饼就火炙烤之用，以小青竹制作最佳，因竹与火接触会产生清香味，有益于茶味，也可用铁、铜制作。

  Bamboo nipper is used for tea cake baking. The best was made of green bamboo as it gave out a delicate fragrance when bamboo meet fire, which was good for the smell of tea. It could also be made of iron and copper.

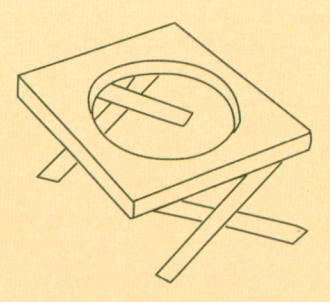

- 交床，将镬固定在风炉上的架子。

  Kettle stand is a stand to fix the cauldron.

- 火夹，以铁或熟铜制作而成，又叫"箸"，用来夹取风炉中使用的木炭。

  Fire tongs are made of iron or processed copper, also called *Zhu*, to take coal for stove.

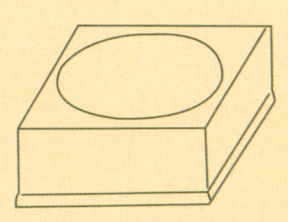

- 碾，碾茶饼为茶末的器具，可用金、银、石、瓷或木等材料制作。
  Tea roller is used to grind tea cake into powder, made of gold, silver, stone or porcelain.

- 都篮，可贮放全部茶具的容器。
  Tea wares container is used to store all the wares.

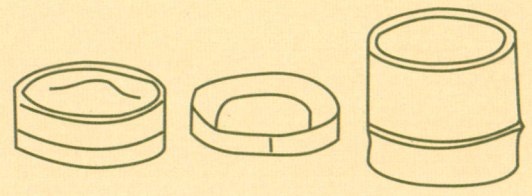

- 罗、盒，以罗筛茶粉，以盒承装用罗筛过的茶末。
  Tea sieve is used to filter the tea powder. Tea case is used to take this powder.

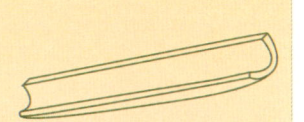

- 揭，取盐器具。
  Ladle is used to take salt.

- 熟盂，贮放第二沸时舀出来的水，以备在第三沸时重新注入茶釜内。
  Hot water jar is used to keep water boiling for the second time, which will be pour back when water boils for the third time.

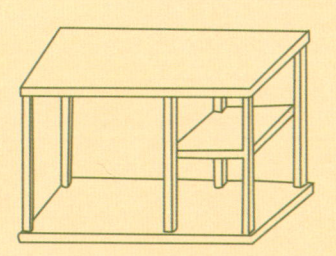

- 具列，盛放各种茶具的架子。
  Teaware shelves are used to store tea wares.

021

茶俗与茶具的发展

The Development of Tea Customs and Teawares

- 碗，饮茶器，以越窑茶瓯为最佳。
  Tea bowl is used for drinking tea. The best tea bowl is from *Yue* Kiln.

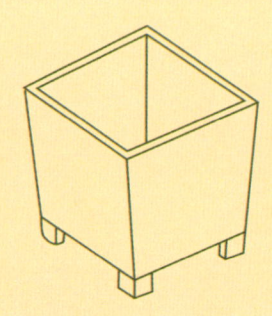

- 涤方，盛放洗涤用水的器具。
  Rinsing container is used to store water for washing.

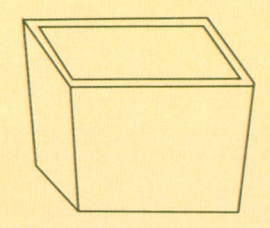

- 滓方，盛放茶渣的用具。
  Dregs container is used to hold tea dregs.

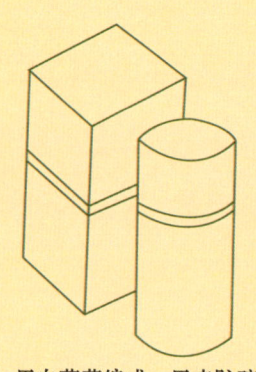

- 畚，用白蒲草编成，用来贮碗。
  Tea bowl container, made of the leaves of cattail is used to hold the tea bowl.

- 鹾簋，盛盐花的容器。
  Salt bottle is used to store salt.

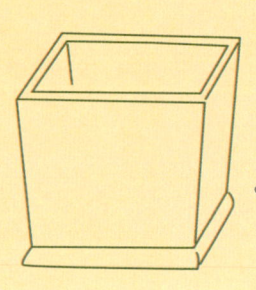

- 水方，盛水容器。
  Water container is used to store water.

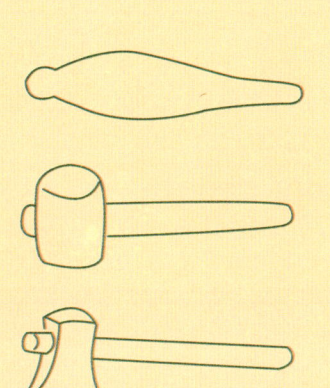

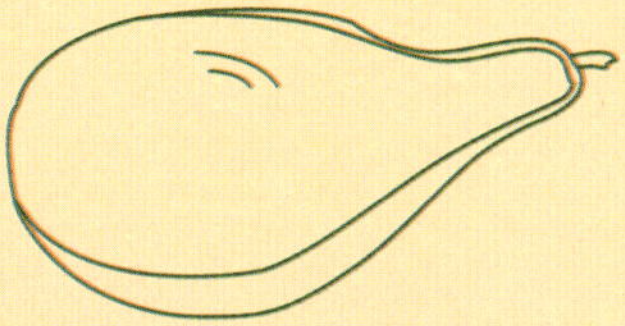

- 瓢，盛水器具，可用葫芦剖制。
  Gourd ladle is used to take water, made of gourd.

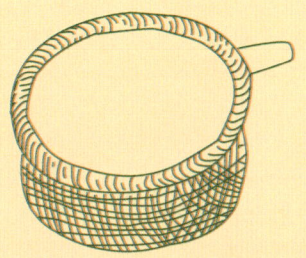

- 炭挝，六角形铁棒，也可制成锤状或斧状，供敲碎木炭之用。
  Tanzhua, coal breaker, is usually a hexagon iron stick, and it can also be made in shape of hammer or axe.

- 漉水囊，过滤水的用具。
  Water filter is used to filter water.

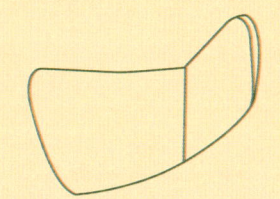

- 纸囊，用来包装烤好的茶饼。
  Paper bag is used to wrap the baked tea cake.

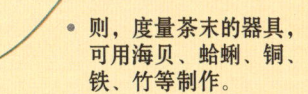

- 则，度量茶末的器具，可用海贝、蛤蜊、铜、铁、竹等制作。
  Tea weighing scoop is used to weigh the tea powder, made of seashells, clams, copper, iron or bamboo.

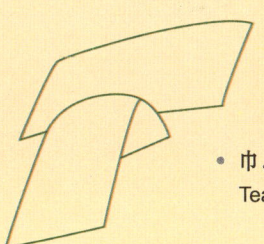

- 巾，用来擦拭清洁器具的布片。
  Tea scarf is used to wipe tea wares.

唐代绘画中的茶具
Teaware in Paintings of the Tang Dynasty

唐代绘画作品留存至今的很少,然而就在这些珍贵的作品中,有不少反映了唐代人饮茶的习俗。通过这些生动的画作可以了解唐代人煮茶、饮茶的习惯以及唐代茶具的诸多种类。

There is only a small amount of paintings of the Tang Dynasty left till now. However, many of these precious paintings reflect tea customs of people in the Tang Dynasty (618-907). Through these vivid paintings we can learn the customs of making and drinking tea and various types of tewares of people in the Tang Dynasty.

唐 阎立本《萧翼赚兰亭图》

这幅画传为唐代著名画家阎立本(约601—673)所作,是迄今为止发现的最早的茶画。画面描绘了儒生与僧人共同品茗的场景,左侧两僧一儒,似一边在谈佛论经,一边在等待香茶奉上。值得注意的是画面左下角一老一少两个侍者正在煮茶调茗的场景。画中布置了一组唐人煮茶的茶具,地上放着茶床,也就是陆羽《茶经》中提到的具列,其作用是摆放茶具。茶床上放着茶碾、茶盏托和一盖罐。茶床边有一老者坐于藤编垫子上,前面放着一茶炉,上置茶铫,正用心煮茶。老者手执茶夹搅动茶铫中刚刚放下的茶末,一旁童子正弯腰捧碗以待。这是极为典型的唐代寺院茶事礼仪图,是唐人茶事的传神写照。

- 唐 阎立本《萧翼赚兰亭图》
Xiao Yi Obtaining the Preface to the Poems Composed at the Orchid Pavilion with a Plot by Yan Liben (Tang Dynasty, 618-907)

Yan Liben, *Xiao Yi Obtaining the Preface to the Poems Composed at the Orchid Pavilion with a Plot*, Tang Dynasty

This painting by Yan Liben (approx. 601-673), a famous painter in the Tang Dynasty, is the earliest painting about tea that has been discovered. It depicts the scene that a scholar is having tea with monks. On the left side of the painting, two monks and one scholar seem to talk about sutra while waiting for the tea. At the left bottom corner of the painting, an old and a young servant are making tea. There is a whole set of teawares in the painting. The tea bed which is mentioned in *The Classic of Tea* by Lu Yu, is on the floor for placing tea wares. There is a tea roller, a tea bowl base and a jar on the tea bed. Next to it an old man is sitting on a vine-made cushion. Concentrating on making tea with a stove in front of him and a tea pan on the stove, he is stirring the tea powder which is just put in the tea pan with tea nippers. A boy is bowing to serve tea with a bowl in his hands. This scene shows a typical tea custom in the temple and a vivid reflection of tea making in the Tang Dynasty.

唐 佚名《宫乐图》

这幅传世名作描绘了唐代宫廷仕女聚会饮茶的场面。宫室中设一张豪华的漆长案，三面围坐仕女十人，皆着华丽衫裙，姿态万千。有的在演奏乐器，有的手执纨扇听曲入迷，也有的手端茶碗忘了品饮，还有的左手执长柄勺从案上的大盆中舀茶，显得雍容典雅，悠闲自得。宴乐饮茶的场面被描绘得精致华丽，这是典型的宫廷仕女自娱茶宴，也反映了茶与音乐相融合的场景。

A Happy Banquet in Palace, Anonymous, Tang Dynasty (618-907)

This famous painting depicts the scene that court ladies in the palace gather and have tea. A

- 唐 佚名《宫乐图》
  A Happy Banquet in Palace, Anonymous (Tang Dynasty, 618-907)

luxurious painted table in the palace is surrounded from three sides by ten court ladies with gorgeous clothes and various gestures. Some of them are playing musical instruments; some listening to the songs with fan in hands; some forgetting to have tea but still holding tea bowls. Some others look elegant and leisure to take tea with a long-handle scoop from the table, delicate and magnificent. This is a classic party for court ladies in a palace, which reflects a scene of the mixture of tea and music.

唐 周昉《调琴啜茗图卷》

画中描绘了五个女性，其中三个系贵族妇女。一女坐在磐石上，正在调琴，左立一侍女，手拿托盘；另一女坐在圆凳上，背向外，注视着调琴者，作欲饮之态；一女坐在椅子上，袖手听琴，另一侍女捧茶碗立于右边。画中贵族仕女曲眉丰肌，艳丽多姿，衣着色彩雅妍明亮。从画中仕女听琴品茗的姿态也反映出唐代贵族悠闲生活的一个侧面。

The Painting Scroll *Chinese Zither Playing and Tea Drinking* by *Zhou Fang*, Tang Dynasty
There are five women in the picture, three of whom are royal ladies. One sits on the stone and plays Chinese zither, with a maidservant holding a tray on her left. Another woman sits on a round stool with her back to the front. She is looking at the Chinese zither player and about to drink the tea. The third woman sits on the chair and listens to the music, with a maidservant holing a tea bowl on her right. The royal ladies in the painting are described with winding eyebrow, beautiful skin. They all look gorgeous with bright-colored clothes. The scene that the royal ladies are listening to music and drinking tea also reflects that royal class in the Tang Dynasty live a very relaxing life.

- 唐 周昉《调琴啜茗图卷》
  The Painting Scroll *Chinese Zither Playing and Tea Drinking* by Zhou Fang (Tang Dynasty, 618-907)

> 宋代的茶俗与茶具

宋代（960—1279）的品饮方式是最优雅的，也是最讲究的。宋代是一个抑武扬文的时代，对文化相当重视，文人的地位也相对较高。在文人为主导的社会里，饮茶也变得更加注重文化和品位，点茶和斗茶就是宋代最有特色的品饮方式。

宋代的点茶习俗

点茶法其实在晚唐时就已出现，到宋代时，成为了从文人士大夫阶层到民间都十分流行的饮茶习俗与时尚。和唐代的煎茶法不同，宋代的点茶法是将茶叶末放在茶碗里，注入少量沸水调成糊状，然后再注入沸水，或者直接向茶碗中注入沸水，同时用茶筅搅动，茶末上浮，形成粥面。

> Tea Customs and Teawares in the Song Dynasty

The way of having tea in the Song Dynasty (960-1279) was the most elegant and delicate. It encouraged civilian while oppressed the military. People then emphasized culture greatly, so scholars had a relatively high position in the society. In a society led by scholars, drinking tea required more culture and art content. Tea dripping and tea contests were the most special way of having tea in the Song Dynasty.

Tea Dripping Custom in the Song Dynasty

Tea dripping was already in use in the late stage of the Tang Dynasty. In the Song Dynasty, it became a very popular tea custom and fashion among scholars,

宋代，朝廷在地方建立了贡茶制度，地方为挑选贡品，需要一种方法来评定茶叶的品位高下。根据点茶法的特点，民间兴起了斗茶的风气。

斗茶多选在清明节期间，因此时新茶初出，最适合参斗。斗茶的参加者都是饮茶爱好者，多的十几人，少的五六人。斗茶内容包括：斗茶品、斗茶令、茶百戏。

斗茶品以茶"新"为贵，一斗汤色，二斗水痕。首先看茶汤色泽是否鲜白，纯白者为胜，青白、灰

officials and civil people. It was different from tea frying in the Tang Dynasty. Song people put tea powder into bowls firstly, stirred to pasty with a little boiling water. Then they added boiling water again, or poured water directly to the tea bowl. At the same time, they stirred with a tea brush. Later, tea powder would float to the surface and form pasty layer.

In the Song Dynasty, the imperial court set a tea tribute system. Local places needed a measure to judge the quality of tea to pay tribute. Common people gradually contested tea due to the features of tea dripping.

- 清 汪承霈《群仙集祝图》
The painting of Servants in a Tea Contest by Wang Chengpei (Qing Dynasty, 1616-1911)

白、黄白为负。汤色能反映茶的采制技艺，茶汤纯白，表明茶叶肥嫩，制作恰到好处；茶汤偏青，说明蒸茶时火候不足；茶汤泛灰，说明蒸茶火候已过；茶汤泛黄，说明采制茶叶不够及时；茶汤泛红，则说明烘焙过了火候。

其次看茶汤的汤花持续时间长短。宋代主要饮用团饼茶，调制时先将茶饼烤炙后碾细。将筛过的极细的茶粉放入碗中，注以沸水，同时用茶筅快速搅拌击打茶汤，使之

• 青白釉出筋壶（北宋）
Bluish-white Glazed Tea Pot with Raised Ridge Design (Northern Song Dynasty, 960-1127)

Tea contest was always held during the period of *Qingming* (Pure Brightness) as new tea would come out at this time, the best time to have such a contest. Participants were tea lovers, dozens or just a few. The tea contest included: tea quality contest, short tea poem contest and tea acrobatics.

Tea quality contest stressed the freshness of tea. One part was the color of the tea infusion. The other was the water mark. Firstly, people would watch whether the tea infusion was fresh and white. The one with pure white would win. Green white, grey white and yellow white ones would lose. The color of the tea infusion could reflect the technique of collecting tea leaves. Pure white one showed that the tea leaves were very fresh and rich and were produced appropriately. Green tea infusion meant that it needed more time to steam. Grey one explained that tea was over-heated. Yellow tea infusion meant that people did not collect tea leaves in time. Red tea infusion showed the tea was over-baked.

Secondly, one would see how long the spray of tea infusion could last. People in the Song Dynasty mainly had tea cake. When they wanted to have tea, they would grind baked tea cake first.

发泡，泛起汤花，称为"击拂"。如果研碾细腻，点茶、点汤、击拂都恰到好处，汤花就匀细，可以紧咬盏沿，久聚不散，这种最佳效果叫做"咬盏"。若汤花不能咬盏，而是很快散开，汤与盏相接的地方

Then they filtered tea powder, put them in tea bowl and added boiling water. At the same time, people used a tea brush to quickly stir the water until it came out with bubbles and spray. This stage was called "hit and stroke". They ground tea cake into delicate powder. At the same time, they dripped tea and water, hit and stroked appropriately. Then the spray would be very fine and smooth. It would stick to the edge of the bowl and stayed there for a long time. This effect was called "grip bowl". If the spray could not stick to the edge but spread outside, then there would be "water mark" between

- 褐釉茶碗（北宋）
  Brown-glazed Tea Bowl (Northern Song Dynasty, 960-1127)

- 粉青釉莲瓣瓷盏托（北宋）
  Pink and Green Glazed Porcelain Bowl Base with Lotus Decoration (Northern Song Dynasty, 960-1127)

- 吉州窑玳瑁釉执壶（南宋）
  Jizhou Kiln Hawksbill Turtle Glazed Ewer (Southern Song Dynasty, 1127-1279)

立即露出"水痕",这就输定了。水痕出现的早晚,是判断茶汤优劣的依据。

斗茶令,即古人在斗茶时,轮流讲故事或吟诗作赋,内容皆与茶有关,如同行酒令,用以助兴增趣。而茶百戏,又称"汤戏"或"分茶",是宋代流行的一种茶道。是将煮好的茶注入茶碗中的技巧,其能使茶汤汤花在瞬间显示出瑰丽多变的景象。汤花或如山水云雾,或像花鸟鱼虫,好似一幅幅水墨图画,这需要较高的沏茶技艺。

water and bowl. This would lose the contest. The time when people found out water mark was the measure to judge the quality of tea infusion.

Short tea poem contest means that during the tea contest, people would tell stories, make poems about tea in turn. This activity shares some features of the drinker's wager game, in order to add to the fun. "tea acrobatics" was also called "infusion show" or "tea dividing". It was a popular tea art in the Song Dynasty. It refers to the technique of pouring tea into tea bowls in order to make tea sprays to show various fabulous pictures in a second. Tea spray can be like mountains, lakes, or cloud and fog. It can also be like flowers, birds, fishes or insects. If it looks like ink paintings, skills are required.

龙凤团茶

　　北宋初期的太平兴国三年（978年），宋太宗遣使至建安北苑（今福建建瓯市东峰镇），监督制造一种皇室专用的茶饼，因茶饼上印有龙凤纹饰，故称为"龙凤团茶"。皇室专用的龙凤团茶表面的花纹是用纯金镂刻而成。龙凤团茶的制作方法十分讲究，采取极细的茶树嫩芽，经过烘焙，研成粉末，加上配料，用特制的木模压成饼状或窝头状，再用印着龙凤图案的细绵纸包装上蜡。饮用的时候打开包装，掰下一点，放进黑釉茶盏，研碎后冲入滚水，待泡沫消失即可饮用。由于团茶的原料和制作都有特殊要求，产量很少，价格也就很高，几乎与黄金等价。

Loong and Phoenix Lump Tea

In the third year of Taipingxingguo (978), an early stage of the Northern Song Dynasty, the second emperor of the Song Dynasty dispatched an envoy to north of Jian'an (now Dongfeng County, Jian'ou City, Fujian Province) to monitor the production of a special tea cake for the royal family. This tea cake was called "Loong and Phoenix Lump Tea" because it had loong and phoenix patterns on its surface. This lump tea was only made for the royal family. The pattern on it was engraved with gold. The process of making this tea was very complicated. People needed to collect extremely tiny tender shoots of tea, bake, grind into powder and add with other seasonings. Then it would be pressed into flat-cake or steamed bun shape with special mould. At last, it would be waxed with delicate cotton paper cover of loong and phoenix patterns. When people wanted to have tea, they would open the cover, break some off and put it in the black-glazed tea bowl. Then they would grind the tea into powder and pour boiling water into the bowl. After the foam was gone, people could take it. As lump tea had special requirements for raw materials and production, there was only a small number of it. Thus the price was very high, almost equal to gold.

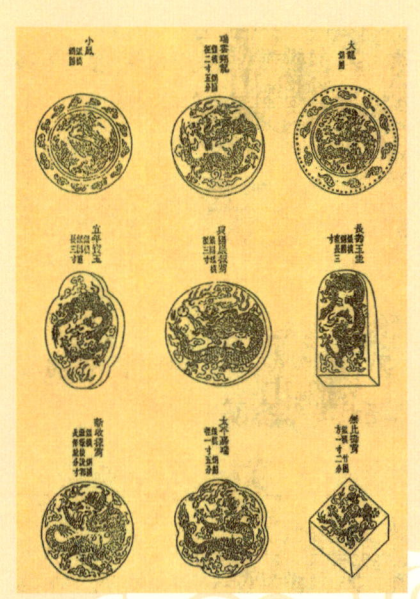

- 龙凤团茶线描图

  A Line Drawing of the Loong and Phoenix Lump Tea

宋代五大名窑与茶具
Five Famous Kilns and Teawares in the Song Dynasty

宋代的瓷器生产以汝窑、官窑、哥窑、钧窑、定窑五个窑口最为有名，后人统称其为"宋代五大名窑"。

Ru Kiln, Official Kiln, *Ge* Kiln, *Jun* Kiln and *Ding* Kiln were the most well-known kilns to produce porcelains in the Song Dynasty. Later generations called them "five famous kilns in the Song Dynasty (960-1279)".

汝窑

汝窑位于古代汝州境内（今河南省临汝县），是北宋徽宗大观元年（1107年）创建的御制官窑。汝窑属于青瓷窑，受越窑影响较大，所烧制的瓷器，不论是胎骨、瓷釉，还是在制作方面都非常精细、规整。汝窑青瓷胎质细腻，胎土中含有微量的铜，所以迎光照看可见微微的红色，还有的胎色灰中略带着黄色，俗称"香灰胎"。而釉色大多呈天青、粉青、天蓝色，也有豆绿、青绿、月白、橘皮纹等釉色。釉面滋润柔和，纯净如玉，纹片晶莹多变，使人赏心悦目。

• 汝窑青釉盏托（宋）
Green-glazed Bowl and Base from *Ru* Kiln (Song Dynasty, 960-1279)

因汝窑曾为宋代宫廷生产大量的生活器具及陈设器，茶具作为皇室日常生活用品之一，也曾被大量生产，但是由于汝窑生产时间短，不久就因战争而停烧，因此存留下来的汝窑茶具非常少，也因而弥足珍贵。

• 汝窑盏托（宋）
Bowl and Base from *Ru* Kiln (Song Dynasty, 960-1279)

Ru Kiln

Ru Kiln, located in ancient Ruzhou (now Linru County, Henan Province), was an official kiln for the royal family created in the first year of Daguan Period in the Northern Song Dynasty (1107). *Ru* Kiln was a celadon kiln and was deeply influenced by *Yue* Kiln. All its porcelains were featured with delicate and neat body, glaze and production methods. Celadon from *Ru* Kiln had very delicate body. As the clay it used contained a small amount of copper, porcelain would show a tiny red when went against sunshine. Some clay contained some grey and yellow color, which was also called "incense ash base". Most of the glaze colors were sky green, light greenish blue, sky blue, as well as bean green, dark green, bluish white and orange peel pattern. The surface of the glaze was soft and gentle, clear like jade. The decoration patterns are also crystal and changeful, which was delightful for eyes. As *Ru* Kiln produced a large sum of daily utensils and exhibits for Song royal families, Teawares were once produced in a large number for daily use of the royal family. However, *Ru* Kiln did not work for a long time. Later, it stopped its production because of the war. Therefore, there were only a few teawares from Ru Kiln, which made these teawares very valuable.

哥窑

哥窑确切的窑场至今尚未发现。据传说为制瓷的兄弟二人各建一窑，哥哥建的窑称为"哥窑"，弟弟建的窑称为"弟窑"。哥窑的主要特征是釉面有大大小小不规则的开裂纹片，俗称"开片"。因小纹片的纹理呈金黄色，大纹片的纹理呈铁黑色，故有"金丝铁线"之说。哥窑瓷器的胎色有黑、深灰、浅灰及土黄多种，而釉色以灰青为主，均质地优良，做工精细，全为宫廷用瓷的式样。传世哥窑瓷器中也有少量的茶具制品，这些茶具因此显得特别珍贵。

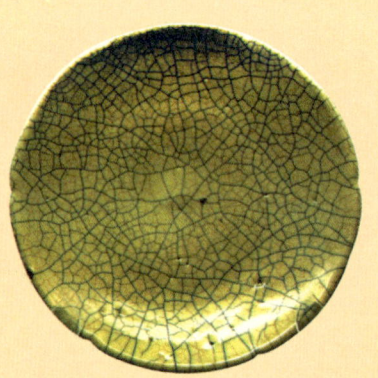

哥窑葵口盘（南宋）
Sunflower-mouth Plate from *Ge* Kiln
(Southern Song Dynasty, 1127-1279)

Ge Kiln

The exact location of *Ge* Kiln hasn't been discovered yet. It is said that two brothers built two kilns separately to make porcelains. The elder brother named his kiln as *Ge* Kiln ("Elder Brother Kiln" in Chinese) and the younger brother named his as "*Di* Kiln" ("Younger Brother Kiln"in Chinese). The main feature of *Ge* Kiln products was the different sizes of cracked patterns on the glaze, also known as "crackle". As the small pattern was gold yellow while the large pattern was iron black, this glaze was also called "gold thread and iron line" pattern. *Ge* Kiln porcelains have different colors like black, dark grey, light grey and yellowish brown. Glaze is mainly grey green, with high quality, delicate work and royal palace porcelain pattern. There are only a few teawares in *Ge* Kiln porcelains that have been kept till now. Thus these teawares are very valuable.

官窑

官窑是宋徽宗政和年间（1111—1117）在京都汴梁建造的，窑址至今没有发现。官窑主要烧制青瓷，器物造型往往带有雍容典雅的宫廷风格。其烧瓷原料的选用和釉色的调配也甚为讲究，釉色以月色、粉青、大绿三种颜色最为流行。釉层普遍肥厚，釉面多有开片。这种开片与同期的哥窑有很大不同，一般来说，官窑釉厚者开大块冰裂纹，釉较薄者开小片；哥窑则以细碎的鱼子纹最为常见。北宋官窑瓷器传世很少，故十分珍稀名贵。

Official Kiln

The Official Kiln was built in Bianliang, the capital city during the Zhenghe Period of the Song Dynasty (1111-1117). The location of this kiln hasn't been discovered yet. The Official Kiln mainly produced celadon and its items always had an elegant palace style. They also focused on the raw materials as well as the mixing of glaze. The three most popular glaze colors were bluish white, light greenish blue and bright green. The glazed layer was always very thick and with crackles. The crackles on Official Kiln products varied a lot from those on *Ge* Kiln products. Generally speaking, thick glaze *Ge* Kiln porcelains had large ice-cracked crackles while thin glaze layer had small crackles. *Ge* Kiln porcelains always had tiny fish egg patterns. Porcelains from the Official Kiln of the Northern Song Dynasty were very precious as there were not many left.

- **修内司官窑盏托（南宋）**

  修内司官窑是北宋灭亡、皇室南迁后，南宋朝廷在都城临安的修内司（今浙江省杭州市凤凰山）建立的第一座官窑。

  Bowl Base from *Xiuneisi* Official Kiln (Southern Song Dynasty, 1127-1279)

  Xiuneisi Official Kiln was the first official kiln built by the imperial court of the Southern Song Dynasty after the Northern Song Dynasty was eliminated and moved to the South. It was located in Xiuneisi, the capital city Lin'an (now Phoenix Mountain, Hangzhou, Zhejiang Province).

钧窑

钧窑是宋徽宗在位时继汝窑之后建立的第二座官窑，广泛分布于钧州（今河南省禹州市）境内，故名"钧窑"。钧窑瓷器都为两次烧成，第一次素烧，出窑后施釉彩，二次再烧。钧瓷的釉色千变万化，红、蓝、青、白、紫交相融汇，灿若云霞。这是因为在烧制过程中，配料掺入铜的气化物造成的艺术效果，此为中国制瓷史上的一大发明，称为"窑变"。

Jun Kiln

Jun Kiln was the second official kiln built during the reign of Emperor Huizong of the Song Dynasty after the Official Kiln, located in Junzhou (now Yuzhou City, Henan Province). Thus it was named as "*Jun* Kiln". All the *Jun* Kiln porcelains were fired twice. Items were fired nakedly for the first time and glazed to fire for the second time. *Jun* porcelains have various kinds of glaze colors. Red, blue, green, white and purple mixed together like rosy clouds. The reason of this art effect was that the ingredients added during firing contained gasified copper. This was a great invention in Chinese porcelain history, called "change in the kiln".

定窑

定窑为民窑，以烧白瓷为主，瓷质细腻，质薄有光，釉色润泽如玉。除烧白釉外，定窑还兼烧黑釉、绿釉和酱釉。造型以盘、碗最多，其次是梅瓶、枕、盒等。在出土的定窑瓷片中，发现刻有"官""尚食局"等字样，这说明定窑的一部分产品是为官府和宫廷烧造的。在传世或出土的定窑器具中，茶具也不少，主要以碗、盏、盏托以及执壶为主，其中紫定盏托以及黑定敞口碗是标准的茶具。

Ding Kiln

Ding Kiln was a civil kiln and mainly produced white porcelains. The quality of its porcelain was fine and smooth, thin and bright and had a glaze that looked like jade. Besides the white glaze, *Ding* Kiln also made black, green and caramel glazes. Most of them were plates, bowls and some plum vases, pillows and boxes. On the pieces of *Ding* Kiln porcelains, people found characters like "官"(official) or "尚食局"(Food Department). This meant that parts of *Ding* Kiln porcelains were made for government and the royal palace. Among the *Ding* Kiln porcelains that have been passed down or discovered, teawares account for a part. They are mainly bowls, bowl bases and pots with a handle to carry. Purple bowl base and black open bowls are standard teawares.

• 钧窑盏托（宋）
Bowl and Base from *Jun* Kiln (Song Dynasty, 960-1279)

• 定窑白釉提梁壶（北宋）
White-glazed Tea Pot with Loop Handle from *Ding* Kiln (Northern Song Dynasty, 960-1127)

• 定窑柿釉盏托（宋）
Persimmon Bowl and Base from *Ding* Kiln (Song Dynasty, 960-1279)

宋代主要茶具类型

与点茶为主的品饮方式相对应，宋代的代表性茶具主要有汤瓶、茶筅和茶盏，三者是斗茶必备的用具。

- 影青瓜棱形汤瓶（宋）
Tea Pot with Green Shadow and Melon Edges (Song Dynasty, 960-1279)

- 黑釉罐（宋）
Black-glazed Jar (Song Dynasty, 960-1279)

Main Types of Teawares in the Song Dynasty

With tea dripping as the major way of drinking tea, typical teawares in the Song Dynasty (960-1279) were the water pot, the tea brush and tea bowl, which were essential in tea contest.

Tea pot, the ware to pour tea, has a complicated making process and plays an important role in tea dripping. Tea pot is also called "pouring item", "pouring pot", "pot with a handle at side" or "pot with handle". Tea pots usually had a big mouth, slipping shoulders, curved front, flat bottom or round foot, with currents on shoulders and belly and handles within the body part. Pot with handles were very common in the Tang Dynasty (618-907), but were mainly for drinking spirit. It did not become teaware until the Five dynasties (907-960) and the Song Dynasty. As required by the process of tea dripping, pot current and handles were lengthened. Especially in the Song Dynasty, the pot was made even slimmer and longer, with a melon edge shape.

The tea brush, made for mixing tea, looks like eggbeater that modern people use. It was made mainly of bamboo. People collected a bunch of thin bamboo and added a handle to make it.

汤瓶是点茶注汤用具，制作非常考究，在点茶过程中起重大作用。汤瓶，又称"注子""注壶""偏提""执壶"。它的基本造型是敞口、溜肩、弧腹、平底或带圈足，肩腹部安流，腹部间安执柄。执壶在唐代就已多见，但以酒器为多，到了五代及宋代，执壶作为茶具出现。因点茶的需要，壶流与执柄开始加长，特别是壶体在宋代变得更加瘦长，并且常制成瓜棱形。

茶筅是烹茶时用来调茶的工具，形状有点类似现代人用的打蛋器，一般以竹为材料，将细竹丝系为一束，加柄制成。

茶盏，因宋人推崇白色的茶汤，所以宋代特别流行用黑釉盏来点茶。宋代的黑釉盏以建窑为代

Since people in the Song Dynasty liked white tea infusion, black-glazed tea bowls were very popular for tea dripping at that time. Black-glazed tea bowl from *Jian* Kiln was the representative of that kind. This bowl usually has a thick body, with both mouths going inward and outward. Both these two shapes of bowls have thick walls. Deep bowl bottom is good for tea dividing; wide bowl bottom serves well to mix with tea brush. People can hit and stroke the tea without blocking by bowl wall; thick body is good for keeping the tea warm. All these advantages surely made black-glazed bowl the favorite one for the Song royal family. Those wares with characters like "tribute bowls" or "tribute emperor" were made as tribute to the Song royal people. Promoted by black-glazed wares

- 黑釉玳瑁斑兔毫盏（南宋）
Black-glazed Bowl with Hare-fur Pattern and Harksbill Spots (Southern Song Dynasty, 1127-1279)

- 吉州窑褐釉剪纸飞凤纹茶盏（南宋）
Jizhou Kiln Brown-glazed Tea Cup with Paper-cut Flying Phoenix Pattern (Southern Song Dynasty, 1127-1279)

表。建窑黑釉盏一般胎体较厚，从造型上看，以敛口和敞口两种为多，无论是哪种造型，其盏壁都很深。盏底深利于发茶；盏底宽则便于茶筅搅拌时不妨碍用力击拂；胎厚则茶不容易冷却。正因为有这么多优点，建窑黑釉盏理所当然地得到宋皇室的偏爱，其器底有"进盏""贡御"铭文的茶盏都是专门上贡给宋皇室的品种。在建窑黑釉器的带动下，南北方的窑场都出现了制作黑釉瓷的高潮，其中与建窑临近的江西吉州窑也在其影响下生产了大量具有自身特色的黑釉茶具。此外北方的磁州窑、定窑及河南的一些窑场也大量生产黑釉茶盏。

from *Jian* Kiln, kilns in both South and North showed a great hit of making black-glazed porcelains. Among these kilns, *Jizhou* Kiln in Jiangxi Province, which was next to *Jian* Kiln, produced many black-glazed tea wares of their own features following *Jian* Kiln. In addition, *Cizhou* Kiln, *Ding* Kiln in the northern part and some kilns in Henan Province also made a great number of black-glazed tea bowls.

《茶具图赞》中的宋代茶具
Teawares in *Diagrams of Teawares* in the Song Dynasty

宋代著名茶人审安老人所作的《茶具图赞》，总结了宋代最有代表性的茶具类型。审安老人的真名叫董真卿，别号"审安老人"，他对宋代的典型茶具做了详细的分门别类，将每种茶具封以官爵，作诗吟咏，还配上线描的十二件茶具图，反映出宋代文人对饮茶以及茶具的重视。

Diagrams of Teawares was written by Senior Shen'an, a famous tea maker in the Song Dynasty. It concluded the most typical teawares in the Song Dynasty. Senior Shen An's real name is Dong Zhenqing and was also called by the name "Senior Shen'an". He categorized typical teawares in the Song Dynasty in detail. He gave every teaware an official's name, made poems and added the paintings of twelve teawares. This reflected that scholars paid much attention to tea and teawares in the Song Dynasty.

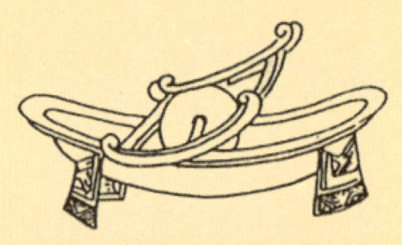

- 金法曹：茶碾，以金属制成，作用是将茶饼碾成茶末。唐代已有茶碾，制作材料不一，宋代沿袭唐代的碾茶法，更为讲究。

  Jin Facao: Tea roller, made of metal, is used for grinding tea cakes into powder. People in the Tang Dynasty had already had tea rollers but they used different materials to make it. People of the Song Dynasty followed the way of grinding tea in the Tang Dynasty but they improved it into a more complicated way.

- 韦鸿胪：茶焙笼，以竹编制而成，竹编时有四方洞眼，所以称为"四窗闲叟"，其最主要的作用是焙茶。因宋代饮用是团饼茶，饼茶加工成型后需存放在干燥的地方以免霉变，宋人于是想出以竹编制茶焙笼的方法来存放茶饼。

  Wei Honglu: Tea baking cage, made of bamboo, has holes all over it so it is also named "relaxed senior with four windows". It's mainly for tea baking. Since people in the Song Dynasty used tea cake lumps, tea cakes had to be stored in dry places to prevent from mildewing after processed. Thus people at that time created this idea of using bamboo cage to store tea cakes.

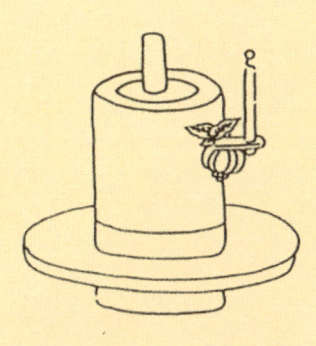

- 石转运：茶磨，因以石头制成，所以被形象地称为"石转运"。用途与金法曹相似，是把茶饼碾碎成粉末状的工具。茶磨有大、小之分，小茶磨适合个人使用，而大的茶磨利用水力等机械装置，基本上由官方来置办。

Shi Zhuanyun: A tea mill, made of stone, is also vividly called "rolling stone transporter". The way of using the tea mill was similar to that of the tea roller, which aimed to grind tea cakes into powder. Tea mills had large and small sizes. The small one was more suitable for personal use while the large one was always set up by the government as it would use mechanical systems like water power.

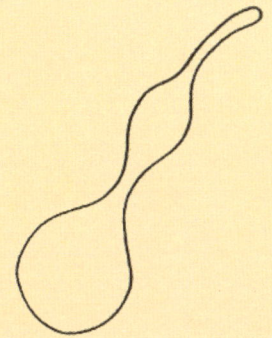

- 胡员外：瓢勺，用以舀水。宋代文人喜欢取江水烹茶，诗人苏东坡就曾在诗中描绘了于月色朦胧的夜晚，用大瓢取江水烹茶饮用的场景。

Hu Yuanwai: It is a gourd ladle for taking water. Scholars of the Song Dynasty loved to use water from river to boil tea. Su Dongpo, a poet, once described a scene that people use big gourd ladle to take water from river for tea at a night with dim moonlight.

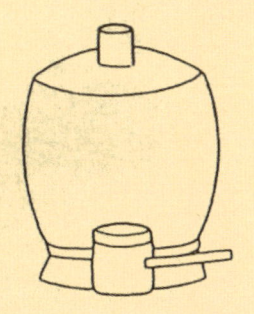

- 木待制：茶槌，用以敲击茶饼，以木材制成。

Mu Daizhi: It is a tea hammer, to hit tea cakes, made of wood.

- 汤提点：水注，又叫"执壶"，主要用途是注汤点茶用。

Tang Tidian: Water container, also called "pot with handle", is mainly used to pour water and drip tea.

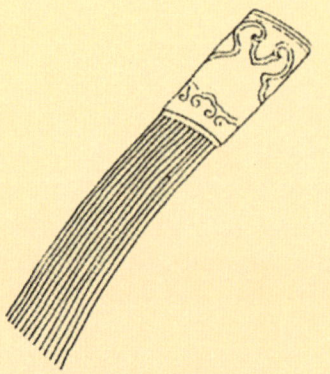

- **竺副帅**：茶筅，以竹子制成，用途是调制茶汤。宋代斗茶取胜的一个关键环节就是以茶筅在盏中调汤时的技巧。

  Zhu Fushuai: Tea brush, made of bamboo, is used for mixing tea. A key point to win tea contest in the Song Dynasty was the technique of mixing tea with tea brush.

- **宗从事**：茶刷，其用途是刷茶末。茶饼碾成茶末、经罗筛选后，就用茶刷扫起集中存放在盒中。

  Zong Congshi: It is a brush for brushing tea powder. After tea was ground and filtered, people would use tea brush to sweep it into a box.

- **陶宝文**：茶盏，这里特指黑釉茶盏，器壁很厚，以兔毫黑釉盏为最佳。

  Tao Baowen: It is a tea bowl, here particularly referred to black-glazed tea bowl. This tea bowl had thick wall. The best one the was the black-glazed bowl with hare fur pattern.

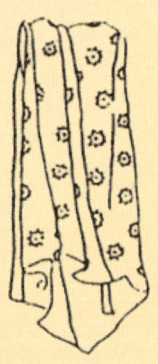

- **司职方**：茶巾，在点茶过程中，保持清洁卫生很重要，因此茶巾是必不可少的。

  Si Zhifang: It is a tea towel. It was vital to keep clean during tea dripping. Thus a tea towel was very essential.

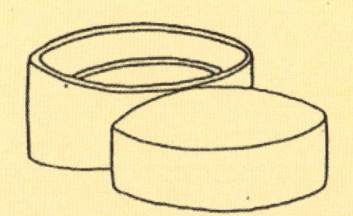

- 罗枢密：罗盒，用途是筛细茶末。茶被碾成茶末后，讲究一点的要过罗筛选。宋代崇尚点茶、斗茶，对茶末的要求极高，如果想在斗茶中取得优势，罗茶也是相当关键的一步。

  Luo Shumi: It is a tea sieve for filtering tea powder. If people paid special attention to the process, tea powder needed filtering with the sieve. People in the Song Dynasty praised tea dripping and tea contest, so they had high requirements for tea powder. If people would like to be in a good position in tea contests, tea powder filtering was also an important stage.

- 漆雕秘阁：盏托，用途是承托茶盏，防止茶盏烫手。宋代盏托形制多样，多以漆制，且以素色漆或雕漆为多。

  Qidiao Mige: It is the bowl base to hold bowl and to prevent it from scalding hands. Bowl bases had various shapes in the Song Dynasty. They were normally made of paint, especially simple color or carving paints.

> 元代的茶俗与茶具

元代的饮茶法可以说是处于从唐宋的以团饼茶为主向明清的散茶瀹泡法的过渡阶段，两种饮茶法都存在，但散茶冲泡已开始兴起。从元墓壁画中我们可以发现茶壶、茶碗、盏托、储茶罐等茶具。

> Tea Customs and Teawares in the Yuan Dynasty

The way of having tea in the Yuan Dynasty (1206-1368) was between tea cake lumps used in the Tang Dynasty (618-907) and soak-and-boil loose tea of the Ming and Qing dynasties (1368-

- 景德镇窑青白釉盏托（元）
  Green and White Glazed Bowl Base from *Jingdezhen* Kiln (Yuan Dynasty, 1206-1368)

元代瓷茶具中，黑釉盏明显减少，而各地方名窑烧制的青瓷和青白釉茶盏、高足杯等茶具数量明显增加。在北方，各地窑口继续大量仿制宋代定窑、钧窑、汝窑、耀州窑等名窑的白瓷和青白瓷茶具；在南方，烧制茶具的名窑则有景德

1911). Although these two ways of having tea existed at the same time, having loose tea began to popularize. We can notice tea pots, tea bowls, bowl bases, and tea storage pots on tomb murals of the Yuan Dynasty.

The number of black-glazed bowls decreased sharply in tea porcelains in the Yuan Dynasty. On the other hand, the number of celadon, green and white glazed tea bowls and cup with high foot made in famous kilns in different places increased greatly. In the northern part of the country, many kilns continued to imitate white porcelain and green and white porcelain teawares made by famous kilns like *Ding* Kiln, *Jun* Kiln, *Ru* Kiln

- 青白釉褐斑葫芦形执壶（元）
  Bluish-white Glazed Calabash-shaped Tea Pot with Brown Speckle (Yuan Dynasty, 1206-1368)

- 卵白釉堆花加彩碗（元）
  Egg-white Glazed Bowl with Flower Pattern and Additional Colors (Yuan Dynasty, 1206-1368)

镇窑、龙泉窑、德化窑等。其中，景德镇窑烧制的青花瓷茶具以其敦朴、精美的造型和装饰工艺，成为茶具史上划时代的杰作。

and *Yaozhou* Kiln in the Song Dynasty. In the South, there were some famous kilns to make teawares, like *Jingdezhen* Kiln, *Longquan* Kiln and *Dehua* Kiln. Because their plain and delicate shape and decoration skills, blue-and-white porcelain teawares made by *Jingdezhen* Kiln were the epoch-making master piece in teawares's history.

- 青花菊花牡丹纹托盏（元）
Tea Bowl and Base with Green Flower, Chrysanthemum and Peony Pattern (Yuan Dynasty, 1206-1368)

宋元时期绘画中的茶具
Teawares in Paintings of the Song and Yuan Dynasties

饮茶的习俗和茶具的使用在宋元时期更加精致和细腻，饮茶已成为宋元时期人们生活的一项重要内容。在这种背景下以茶入画的现象较唐代更加普遍。

Custom of having tea and the use of teawares became more complicate and delicate in the Song and Yuan dynasties. Having tea had become an important part of the daily lives of people in the Song and Yuan dynasties. Under such a circumstance, it was common to take tea as the theme of paintings.

北宋　赵佶《文会图》

宋徽宗赵佶（1082—1135），在位25年，是一位才华出众的风流天子，善书画、山水、人物、花鸟、墨竹无一不精，而且精通茶艺，曾著有《大观茶论》。

- 北宋 赵佶《文会图》
  The Painting of Scholars' Banquet by Zhao Ji (Northern Song Dynasty, 618-907)

《文会图》是公认的描绘茶宴的佳作，展现出宋代文士雅集的典型场景。整个活动在一处宽敞而幽雅的庭院中进行，出席宴会的都是文士或官员，共十一人，个个神采奕奕。

　　画面上共有侍茶、侍酒八人，一名身着官服者，似为宴会的总管。左下角设有茶桌与酒桌，茶桌与风炉设在左边，桌上摆着茶盒和茶碗，一位侍茶正用茶则从茶盒往茶碗中分茶；炉火正旺的燎炉上置两把执壶，两名侍者正在向来宾奉茶；在风炉前方地上还放着都篮，里面整齐地摆放着备用的茶碗；右边的酒桌上放两把酒壶，一个酒坛放在地上；宴会总管已端起一个托盘，盘中似放着几杯斟满酒的小酒杯，好似正在准备开宴。宴会桌之后，花树间设一桌，上置香炉与琴。

　　《文会图》中宴会场面宏大而雅致，故友相逢，三三两两亲切交谈，有的离开自己的座位走到老友跟前。此画将茶宴同酒宴、珍馐、插花、音乐、焚香等融于一图之中，实为宋代茶画之精品。

The painting of Scholars Banquet by Zhao Ji, Northern Song Dynasty

Zhao Ji, Emperor Huizong (1082-1135) of the Song Dynasty was in power for 25 years and was also a talented and romantic emperor. He was good at painting and calligraphy, no matter whether they were paintings about mountain and river, people, flowers and birds or about bamboo. He was also a master of making tea. He wrote *General View of Tea*.

　　The Painting of Scholars' Banquet was a well recognized masterpiece about tea meeting, which reflected a typical scene where scholars of the Song Dynasty gathered together. The whole activity was held in a wide and peaceful yard. There were eleven participants, who were all scholars or officials. They seemed to be very energetic.

　　In the painting, there are eight tea and wine servants. The person wearing official uniform looked like the manager of the banquet. At the left bottom corner of the painting, there were tables for tea and spirit. Tea table was on the left side of the stove, with tea boxes and bowls on it. One servant was dividing tea from tea box into bowls with tea weighing scoop; two pots with handle were on the stove of big fire and two servants were serving tea to the guests; a tea basket was set in front of the stove, with spare tea bowls neatly in it; there were two wine pots on the wine table on the right side and one wine jug on the floor; the manager of the banquet had taken a tray. It seemed that there were some small cups with wine on the tray and the banquet was almost ready. Behind the tables, there was another one in flowers and trees with censer and traditional Chinese instrument.

　　The scene in the painting of scholars' banquet was grand yet delicate: old friends met again; people talked to one and another; some one left his seat and came to an old friend. This painting was a master piece in paintings about tea in the Song Dynasty, which perfectly compromized the tea banquet, wine banquet, delicacies, flowers, music and incense.

南宋 刘松年《撵茶图》

刘松年（约1155—1218），钱塘（今杭州）人，其传世画作虽极少，却有三幅是茶画。

宋代饼茶饮用前需将茶饼碾成粉末，过筛后冲点，《撵茶图》生动地再现了碾茶这个工序。画面左边左上角是几株高大的芭蕉，树下方桌旁的侍者一手执茶瓶注汤于茶盏中，茶瓯边放着点茶用的茶筅，另一手持茶盏，桌上另一边放着茶盏及其他茶具。方桌前方是茶炉，茶炉上放着茶铫。此侍者身后是贮水的泉缶，圆肚，上覆一箬叶。画面左下角一侍者坐于矮几上碾茶，神情专注，碾茶的工具是茶磨，石磨上挂着小茶巾，前面还有茶匙和茶刷。画面右侧一僧伏案执笔，一人坐其旁，另一人与僧相对而坐。《撵茶图》可见碾茶、煮茶等茶事活动已经与文人的笔墨生活融为一体，成为文人生活的一部分。

The Painting of Grinding Tea by Liu Songnian, Southern Song Dynasty
Liu Songnian (approx. 1155-1218) from Qiantang (now Hangzhou), left only a few paintings. However, three of them are about tea.

In the Song Dynasty, people ground tea cakes into powder, filtered and then poured water into it. *The Painting of Grinding Tea* vividly reflected this process. At the top left corner, there were several large banana trees, under which, a servant was pouring water into a tea bowl with the tea pot next to the table. The tea brush was next to the tea bowls. The other hand was holding

- 南宋 刘松年《撵茶图》
  The Painting of Grinding Tea by Liu Songnia (Southern Song Dynasty, 1127-1279)

the tea bowl and there were tea bowls and other tea wares on the table. The stove, with a pan on it, was in the front of the square table. A fountain jar, at the back of this servant, had a round belly and an indocalamus leaf on top. At the bottom left corner of this painting, one servant was concentrating on grinding tea on a small low table. He used tea roller to work. There was a small tea towel on the stone roller with tea scoop and brush in the front. On the left side, a monk was writing on the table with one person next to him and another one in front of him. Through *The Painting of Grinding Tea*, we could see that tea making activities, like grinding or boiling tea went well with writing and other arts. It had became part of the scholars' life.

南宋 钱选《卢仝煮茶图》

钱选（1239—1302），字舜举，号玉潭，浙江吴兴（今湖州）人，宋末元初时期著名的画家。这幅《卢仝煮茶图》的画面中部绘有高大的太湖石和深绿色的芭蕉树。下部是体现主题的烹茶场面，主人公卢仝身着白衣、纱帽，美髯飘扬，气定神闲，坐于山岗平石上。卢仝身边伫立着来送茶的人。风炉前烹茶的红衣仆人，右手持扇向炉口扇风。主人、差人、仆人三者同现于画面，三人的目光都投向茶炉，表现了卢仝得到好茶后迫不及待烹饮的心情。此画抒发了作者闲逸的心境，在平静如水的线条之中，展现出遁世隐居、洁身自好的高雅情趣。

The Painting of Lu Tong Boiling Tea by Qian Xuan, Southern Song Dynasty

Qian Xuan (1239-1302), also named Shunju or Yutan, was from Wuxing (now Huzhou), Zhejiang Province. He was a famous painter of the late Song and early Yuan dynasties. In the middle of *The painting of Lu Tong Boiling Tea*, there were some huge Taihu stones and dark-green banana trees. The bottom part reflected the scene of boiling tea. The main character Lu Tong wore white cloth, gauze cap with beautiful beard. He sat on the flat stone of the hill and looked handsome and relaxed. The person standing next to him was serving tea. The servant in red was boiling tea in front of the stove and making wind with a fan in his right hand. Master, errand boy and servant were all in the scene and they were all

• 南宋 钱选《卢仝煮茶图》
The Painting of Lu Tong Boiling Tea by Qian Xuan (Southern Song Dynasty, 1127-1279)

looking to the stove, which showed that Lu Tong couldn't wait to drink after getting high-quality tea. This painting expressed a relaxed mood of the painter. He described a noble interest of seclusion and kept himself pure by using peaceful lines in the painting.

辽 河北宣化辽墓室壁画

1971年，河北宣化郊外下八里村相继出土了数座辽代墓葬，墓室内彩色壁画和出土器物十分丰富，其中有多幅反映不同茶事场面的壁画，包括点茶图、为点茶做准备工作的煮汤图、妇人饮茶听曲的娱乐场面图、向饮茶者进茶场面图和茶作坊中碾茶、煮点、筛选等一系列工序图等等，从这些壁画中，可以较完整地看到当时人们品饮艺术的发展和变化。

Murals in the Tomb of the Liao Dynasty in Xuanhua, Hebei Province

In 1971, several tombs of the Liao Dynasty were excavated in Xiabali Village, outside Xuanhua, Hebei Province. There were various kinds of colorful murals and wares from these tombs. Among these murals, some reflected different scenes of making tea, including the painting of tea dripping, water boiling to prepare for tea dripping, entertainment scene that women were listening to music while having tea, serving tea to the drinker and tea making processes in the tea workshops like tea grinding, tea dripping and filtering. These murals reflected the development and evolution of drinking art at that time.

河北宣化辽墓壁画《将进茶图》

画面正中绘一赭色桌子，桌上有红色托子、白色盏碗、红色箱子和白色深腹盆。桌前摆一五足火盆，盆内有火炭和白色瓜棱执壶。画面右侧那人双手端一白色唾盂，目视火盆。桌后那人双手捧盏碗，小心翼翼，而手持团扇者好像正叮嘱她别烫着。整个画面反映的是点好茶后准备向饮茶者进献的场面。

The Painting of Urging People to Have Tea from Mural in the Tomb of the Liao Dynasty, Xuanhua, Hebei Province

In the middle of the painting there was a reddish brown table with red base and white bowls, red box and white deep basin on top. A five-foot fire brazier was in the front holding charcoal, alongside a white melon-wedged pot with handle. On the right side, a person was holding a white spittoon and looking at the brazier. The one behind the table was carefully holding the bowl while the one with a fan seemed to warn her from being scalded. The whole painting reflected the scene that people were about to serve tea after tea dripping.

- **河北宣化辽墓壁画《茶道图》**

  这幅图表现的是一个有趣的生活场景：在饮茶之前，一个女孩登在一个男孩的肩上，从在屋顶吊着的竹篮里往外取桃子。画面中除了女主人正在注视着茶瓶里将要煮好的茶汤之外，藏在柜子后面的四个儿童都看着那个女孩子，而下边一个仆人正在提起围裙接桃子。这幅富有生活情趣的《茶道图》说明当时人们在饮茶的同时还有吃水果的习惯。

The Painting of Making Tea from Mural in the Tomb of the Liao Dynasty, Xuanhua, Hebei Province

This painting depicted an interesting life scene: before drinking tea, a girl is standing on the shoulders of a boy in order to get a peach from the bamboo basket on the roof. In this painting, except the hostess was watching the tea which was about to ready in the pot, all four children hiding behind of the closet were looking at the girl. The servant at the bottom was raising his apron to hold peaches. This interesting and joyful *Painting of Making Tea* showed that people had the custom of eating fruit while having tea.

- **河北宣化辽墓壁画《煮汤图》**

  画面右侧绘一张红色长方形高桌，桌上放白色盏碗、盏盖、托子、花口盘和四层的波罗子（一种多层套盒，每一层又分为若干小格，用来盛装茶点）。桌前置一三足火盆，盆中有燃烧的火炭和一个长流瓜棱执壶，一儿童双手执扇用力扇火。桌后二人，右边那人左手拿白色平底大盘，盘上有两只白色茶杯，与左边那人低声交谈，二人好像在等待执壶中的水烧开。整个画面是一幅为点茶做准备工作的煮汤场景。

  The Painting of Boiling Water from Mural in the Tomb of the Liao Dynasty, Xuanhua, Hebei Province

  On the left side of the painting, there was a red square high table with white bowls, lids, bowl bases, fancy top plates and a *Boluozi* (a box with several layers and each layer was divided into several parts to take tea cakes). In front of the table, some burning charcoal and a long and slim pot with handle were in a three-foot brazier. A boy made an effort to pan the fire with his hands. There were two people behind the table and the one on the right held a big white flat plate with two white tea cups on it. He was talking softly with the one on the left. They seemed to be waiting for the water in the pot to boil. The whole scene depicted preparing for tea dripping.

元 赵孟頫《斗茶图》

宋代盛行"斗茶"，赵孟頫作的《斗茶图》是反映"斗茶"活动最为著名的画作之一。赵孟頫（1254—1322），字子昂，元代著名的画家，对当时和后世的画风影响很大。这幅《斗茶图》上有四人，两人为一组，左右相对，每组中的长髯者为主角，各自身后的年轻人是助手，属配角。左面这组，年轻人执壶注茶，身子前倾，两小手臂向内，两肘部向外挑起，姿态健壮优美有活力。年长者左手持杯，右手拎炭炉，昂首挺胸，面带自信的微笑，好似已是胜券在握。右边一组，年长者左手持已尽之杯，右手将最后一杯茶品尽，并向杯底嗅香，年轻人则在注视对方的目光时将头稍稍昂起，似乎并没有被对方的踌躇满志压倒，大有鹿死谁手还未知的神情。画面结构紧凑、动静结合，人物栩栩如生、鲜明生动，线条十分细腻。斗茶比技巧、斗输赢，趣味性很强，茶农、僧人、文士均喜爱，场面十分热闹。元代斗茶之风已渐渐消隐，赵孟頫的这幅《斗茶图》是画家对宋代斗茶活动的追想与怀念。

The Painting of Tea Contest by Zhao Mengfu, Yuan Dynasty

Tea contest was very popular in the Song Dynasty (960-1279). *The Painting of Tea Contest* by Zhao Mengfu is one of the most famous paintings reflecting tea contest. Zhao Mengfu (1254-1322), also named Zi'ang, was a well-known painter in the Yuan Dynasty (1206-1368) and had great influence on the style of paintings at that time and later. In this painting, there were four people. Two people sat opposite of each other. The two people with beard were the main characters and the younger ones behind them were assistants, or supporting characters. In the group on the left, the young one was pouring tea with a pot. He leaned forward a little bit, with his arms heading inwards and elbows outwards. He looked strong, handsome and energetic. The senior one held a cup in his left hand and a stove in his right hand. He stood with his head high and chest out and gave confident smile. It seemed that he knew he would win. In the group on the right side, the senior one was holding an empty cup with his left hand. He was finishing the last cup of tea on his right hand and looked like smelling the fragrance on the bottom of the cup. The younger assistant raised his head when watching the other side. It seemed that he was not frightened by the confidence of their opponents but thought it was not certain that who would win the game. The whole painting had a compact structure while perfect combing action with steady state. All the characters were vivid, distinctive with delicate lines. The tea contest focused on skills and people would compete for the winner. So it was a very interesting game. Tea farmers, monks and scholars all liked it, and the scenery was very lively. In the Yuan Dynasty, tea contest gradually faded away. Therefore, this *Painting of Tea Contest* by Zhao Mengfu was a recall and yearning of tea contests in the Song Dynasty.

- 元 赵孟頫《斗茶图》

The Painting of Tea Contest by Zhao Mengfu (Yuan Dynasty, 1206-1368)

> 明清时期的茶俗与茶具

宋代的饮茶方式发展到元代已开始走下坡路，因饼茶的加工成本太高，而且其在加工过程中把茶汁榨尽，也违背了茶叶的自然属性。所以到了元代，团饼茶已日渐式微，之前已经出现的散茶从明代（1368—1644）开始流行。

明清时期的散茶

明朝初期，虽然散茶在民间已经逐渐得到普及，但进贡宫廷的贡茶仍然采用福建的团饼茶。后来，明太祖朱元璋认为进贡团饼茶太"重劳民力"，于是下令停止龙凤饼茶的进贡，而改进芽茶（散茶的一种）。明太祖的诏令，对进一步破除团饼茶的传统束缚，促进散茶

> Tea Customs and Teawares in the Ming and Qing Dynasties

The way people in the Song Dynasty drank tea was declining in the Yuan Dynasty. The reason was the high cost of making tea cakes. In addition, people would squeeze all the tea fluid out when making tea but this went against the natural feature of tea leaves. Therefore, in the Yuan Dynasty, people seldom saw tea cake lumps, while loose tea, which showed up earlier, flourished from the Ming Dynasty (1368-1644).

Loose Tea in the Ming and Qing Dynasties

At the first stage of the Ming Dynasty, loose tea was very popular among common people, while tea cake lumps from Fujian Province were used as

的蓬勃发展，起到了有力的推动作用。明朝茶业在技术革新、各种茶类的全面发展以及名茶种类的繁多上形成了自己的时代特色。

明清时期，在茶的生产上，出现了不少新的茶树种植和茶叶加工技术。由于工艺技术的改进，各地名茶的发展也很快，品类日渐繁多。宋代时知名的散茶寥寥无几，到了明代，有名的散茶品种达到了90余种。

tribute. Then Zhu Yuanzhang, the first emperor of the Ming Dynasty thought that tea cake lumps as tribute were too heavy a burden for ordinary people. Thus he stopped using loong and phoenix tea cakes as tribute and changed to tea buds (one kind of loose tea). His order played a positive role in abolishing the constraint of tea cake lumps and promoting the development of loose tea.

Tea industry in the Ming Dynasty developed its own unique features in terms of technique innovation, an overall development of all kinds of tea and enriching the kinds of famous tea. In the Ming and Qing dynasties, there were many new tea trees and tea processing techniques in the process of tea producing. Due to the improvement of skills, famous tea in different places developed fast with more kinds. In the Song Dynasty, well-known loose tea was hard to find. However, in the Ming Dynasty, there were up to 90 kinds of famous loose tea.

清代宫廷中的"三清"茶具

　　清朝历代皇帝都好饮茶，宫廷里饮茶之风很盛行，清宫设有茶库，宫廷内务府专门设有"御茶房"，每年收取进贡名茶30多种。根据清史档案记载，乾隆帝在位的60年间，每年新正（农历新年）必举行茶宴，择吉日在重华宫由乾隆帝亲自主持，茶宴有一套规范的礼仪程序，先是由皇帝出题定韵，出席茶宴的群臣竞相赋诗联句，接下来是品茗、食茶点。茶宴上准备的茶名曰"三清茶"，由梅花、佛手、松实（松子）加融化的雪水烹制。

　　"三清"茶具是清代皇室定制的茶具，有一定的规范，品种包括瓷器、漆器、玉器等，以盖碗为多。一般在茶具的外壁写上乾隆皇帝御制诗："梅花色不妖，佛手香且洁。松实味芳腴，三品殊清绝……"在茶宴上，乾隆皇帝会把"三清"茶具赏赐给一些在赋诗联句中表现突出的大臣，作为对他们的嘉奖。

"Three-clean" Teawares in the Qing Palace

Every emperor in the Qing Dynasty (1616-1911) loved drinking tea, so tea was very popular in the imperial court. The imperial court of the Qing Dynasty established tea storage and specially designed "royal tea house" with 30 kinds of tea tributes every year. According to the record of the history of the Qing Dynasty, Emperor Qianglong held a tea banquet on *Xinzheng* (the new year by the lunar calendar) every year during the 60 years when he was in power. Emperor Qianlong would choose a good day to host the banquet himself in *Chonghua* Palace. The banquet had a standardized etiquette order. Firstly, the emperor would give a question and fix the rhyme. All the participants would compete to make poems. Later, people could drink tea and have tea cakes. The tea people had on this banquet was named "three-purity" tea, which was made from plum blossom, finger citron, *Songshi* (pine nut) boiled with melted snow.

　　The "three-purity" teaware was made particularly for the imperial court of the Qing Dynasty, so there were some requirements. This teaware always included porcelain, lacquer, and jade, most of them covered teacups. On the outside wall of the teawares, there were always the poems composed by Emperor Qianlong: "The plum blossom is not coquettish while finger citron is aromatic and pure. Pine nut is savory and all these three are very pure." During the banquet, Emperor Qianlong would give "three-purity" teawares to those officials who stood out in composing poems as their reward.

- 清宫旧藏茶叶包装盒
  Box to Store Tea in the Imperial Court of the Qing Dynasty

- 清代宫廷举行茶宴时用的茶碗
  Tea Bowls Used in the Banquet in the Imperial Court of the Qing Dynasty

明清时期的特色茶具

明代的散茶种类繁多，散茶不需碾末过罗后冲饮，较前代简便多了，而且还原了茶叶的自然味道。由于茶叶不再碾末冲泡，以前茶具中的碾、磨、罗、筅、汤瓶之类的茶具皆废弃不用，宋代崇尚的黑釉盏也退出了历史舞台，代之而起的是景德镇的白瓷。用雪白的茶盏来衬托青翠的茶叶，可谓尽茶之天趣也。

明清的茶具从材质上来讲，以瓷器为主。明清两代的瓷器主要以景德镇为中心，景德镇成了名副其实的瓷都。在景德镇窑元代青花、釉里红、红釉、蓝釉、影青、枢府釉发展的基础上，经过几代窑工的努力，烧制出了不少创新品种。其中从釉色上来说有青花、釉里红、青花釉里红、单色釉（包括青釉、白釉、红釉、绿釉、黄釉、蓝釉、金彩），仿宋五大名窑器、粉彩、五彩、珐琅彩、斗彩等。

茶壶

茶壶在明清两代得到很大的发展。在此之前，把有流带柄的容器皆称为"汤瓶"或"偏提"，到了

Special Teawares in the Ming and Qing Dynasties

People in the Ming Dynasty (1368-1644) had more kinds of loose tea. Loose tea need not grinding or filtering, so it was more convenient to use than before. It also kept the natural smell of tea. As people needed not to grind tea before drinking, the tea roller, mill, sieve, brush or water pot were no longer in use. Black-glazed bowl which was very popular in the Song Dynasty also retreated from this arena of history. Instead, people began to advocate white porcelain from *Jingdezhen* Kiln. It was the greatest pleasure to drink tea when using snow-white bowl to show the green tea leaves.

From the perspective of materials of teawares in the Ming and Qing dynasties (1368-1911), porcelain was the most popular one. Jingdezhen was the centre of porcelain in the Ming and Qing dynasties and a real capital of porcelain. Following the green porcelain in the Yuan Dynasty (1206-1368), underglaze red porcelain, red-glazed porcelain, blue-glazed porcelain, grisaille porcelain and flower petal porcelain made in *Jingdezhen* Kiln, people created many new types after the effort of several generations.

明代，真正用来泡茶的茶壶才开始出现。壶的使用弥补了盏茶易凉和易落尘的不足，也大大简化了饮茶的程序，受到世人的极力推崇。

虽然有流有柄，但明代用于泡茶的壶与宋代用来点茶的汤瓶还是有很大的区别。明代的茶壶，流与壶口基本齐平，使茶水可以保持在壶内而不致外溢，壶流也制成"S"形，不再如宋代强调的"峻而深"。明代茶壶尚小，以小为贵，因为"壶小则香不涣散，味不耽搁。"清代的茶壶造型继承明代风格，但制作材料上有了很大的改进，瓷茶壶和紫砂壶大量出现。

- 耀州窑青釉壶（明）
Celadon Tea Pot from *Yaozhou* Kiln (Ming Dynasty, 1368-1644)

From the point of glaze color, there were blue, underglaze red, blue-and-white underglaze red, single-color glaze (including blue, white, red, green, yellow and golden color), colors that imitated the five famous kilns in the Song Dynasty; famille rose, polychrome, enamel-color, contending-color, etc.

Tea Pot

Tea pots developed greatly in the Ming and Qing dynasties. Before that, all containers with handles were called "water pot" or "pot with handle". Pot only for tea was in use in the Ming Dynasty. Pot made for the shortage that tea easily went cold in bowl and caught dust. It also simplified the process of having tea, so it was greatly welcomed by people at that time.

Although both with the current shape and handle, the pot for tea in the Ming Dynasty was very different from the water pot for tea dripping in the Song Dynasty. The current shape of the tea pot in the Ming Dynasty was almost even with the pot mouth, so tea would not overflow from it. The shape of tea pot is in an "S" style, instead of a "steep and deep" one which was essential in the tea pot of the Song Dynasty. Tea pot in the Ming Dynasty was

- 三足紫砂圆壶（明）
Purple Clay Round Tea Pot with Three Feet (Ming Dynasty, 1368-1644)

- 粉彩花蝶图茶壶（清）
Famille Rose Tea Pot with Butterfly Pattern (Qing Dynasty, 1616-1911)

盖碗

从茶具形制上讲，除茶壶和茶杯以外，盖碗是清代茶具的一大特色。盖碗一般由盖、碗及托三部分组成，象征着"天、地、人"三才，反映了中国古老的哲学观。盖

- 青花粉彩福寿盖碗（清 嘉庆）
Blue-and-white Famille Rose Tea Bowl with Cover (Jiaqing Period of the Qing Dynasty)

always small. And small ones were the precious ones as "Small tea pot will keep the smell from going away and is good for drinking". The shape of the Qing tea pots followed that in the Ming Dynasty. However, they improved a lot in the material and there were many porcelain tea pots and purple clay tea pots.

Tea Bowl with Fitted Cover

In term of the shape of teawares, besides tea pot and tea cups, tea bowls with fitted cover were another unique feature of teawares in the Qing Dynasty (1616-1911). It included cover, bowl and base, which meant "sky, earth and people" respectively, an ancient Chinese philosophy. The tea bowl with fitted cover could prevent the bowl from

碗的作用之一是防止灰尘落入碗内，起了有效的防尘作用；其二是防烫手，碗下的托可承盏，喝茶时可手托茶盏，避免手被烫伤。明清时期，景德镇生产了大量的陶瓷盖碗，品种包括青花、五彩、斗彩、粉彩、釉里红、单色釉等。

getting dust; also, it protected hands from being scalded. The base could hold the bowl and people could hold the base in order not to be scaled. In the Ming and Qing dynasties, a large number of porcelain tea bowls with fitted cover were produced in Jingdezhen, including celadon, polychrome, clashing color, famille rose, underglaze red and single-color ones.

- 胭脂红釉盖碗（清 乾隆）
Carmine-glazed Tea Bowl with Cover (Qianlong Period of the Qing Dynasty)

- 青花诗文三托盖碗（清）
Blue-and-white Tea Bowl with Engraved Poem (Qing Dynasty, 1616-1911)

茶洗

由于明人饮用的是散茶，在散茶加工过程中可能会沾上尘垢，于是在泡茶之前多了一道程序——洗茶，茶洗就是洗茶的专门茶具。茶洗形状像碗，上下二层，上层底部有许多小孔，茶叶放在碗里用水冲洗，沙砾污垢都随着水流从孔中流出。也有的茶洗做成扁壶式。

Tea Washing Container

Since people in the Ming Dynasty used loose tea which might catch dust during processing, washing tea became another stage before making tea. The tea washing container is a special utensil for washing tea. It looks like a bowl with two layers. At the bottom of the upper one, there are many small holes. People put tea in it and wash with water. Dust and grits

• 各式茶洗
Different Kinds of Tea Washing Containers

贮茶具

明代散茶的流行对茶叶的贮藏提出了更高的要求，炒制好的茶叶如果保藏不善，茶汤的效果就会大打折扣，所以贮茶器具的优质比唐宋时显得更为重要。散茶的保存环境宜温燥，忌冷湿。一般来说，明

will flow out from the holes with water. Some tea washing containers are made into a flat pot.

Tea Storage Utensil

As loose tea became very popular in the Ming Dynasty, it required a higher standard of tea storage. If fried tea

• 青花茶叶罐（明）
Blue-and-white Tea Storage Jar (Ming Dynasty, 1368-1644)

• 紫砂茶叶罐（清 乾隆）
Purple Clay Tea Storage Jar (Qianlong Period of the Qing Dynasty)

• 金属茶叶罐（清）
Metal Caddy (Qing Dynasty, 1616-1911)

代的散茶保藏采用瓷瓶或紫砂瓶。将焙干的茶叶放入茶瓶，再将细竹丝编织的箬叶覆盖其上，而后瓶口用六七层纸封住，上面再压上白木板，放在干净处存放。需要用时，从大瓶中取一些茶叶放入干燥的小瓶待用。出土及传世的明代茶叶瓶、茶叶罐形制各异，大小不一。清代茶叶罐的种类更加丰富多彩，或圆或方，或瓷或锡，造型千姿百态。

茶船

茶船，亦称"茶托子""茶拓子""盏托"，以承茶盏防烫手之

leaves could not be stored properly, its quality would be greatly damaged. Thus, compared with the situation in the Tang and Song dynasties, tea storage utensils were more vital then. A mild and dry environment was the best place to keep loose tea while cold and wet would damage the tea. Generally speaking, people used porcelain pots or purple clay pots to keep loose tea. People would put dried tea leaves into the pot and cover with an indocalamus leaf made of thin bamboo. Then the pot mouth would be covered with six or seven pieces of paper, white board and put in a clean place. When needed, people would take some tea from the big pot into small dry bottle for further use. Tea storage bottle and jars varied a lot in shape among those discovered or passed down from the Ming Dynasty. Tea storage had more kinds in the Qing Dynasty. They had thousands of forms, round or square, made of porcelain or tin.

Tea Boat

Tea boat, also called "tea base", "tea couch" or "bowl base", is made to prevent hands from being scalded. Since it looks like the shape of a boat, it is then named tea boat or tea ship. The tea boat

用。后因其形似舟，遂以茶船或茶舟名之。茶船最初是从盏托演变过来的。明清之际茶船相当流行，形制各异，材料有陶瓷、漆木、银质、锡金属等。

was firstly developed from the bowl base. During the Ming and Qing dynasties, tea boats with different shapes and materials like ceramic, porcelain, lacquer wood, silver or tin were popular.

• 铜胎画珐琅牡丹纹茶船（清）
Enamel-color Copper Tea Boat with Peony Pattern (Qing Dynasty, 1616-1911)

• 松石地印花茶船（清）
Tophus Tea Boat with Engraved Pattern (Qing Dynasty, 1616-1911)

• 青花开光诗句海棠式茶托（清 嘉庆）
Frame Outlining Blue-and-white Tea Base with Begonia Pattern (Jiaqing Period of the Qing Dynasty)

茶壶桶

唐宋之际由于盛行煮茶、点茶，并不存在茶水需要保温的问题。而到了明清两代，散茶投入茶壶中，为了不让茶水过快冷却，有人发明了茶壶桶。茶壶桶看上去就

Tea Pot Bucket

In the Tang and Song dynasties, boiling tea and tea dripping did not have the problem of keeping the temperature of tea. However, in order to prevent tea from cooling quickly in the tea pot, some people invented tea pot bucket in

- 藤编茶壶桶（清）
Tea Pot Bucket Made of Vines (Qing Dynasty, 1616-1911)

- 木制茶壶桶（清）
Wooden Tea Pot Bucket (Qing Dynasty, 1616-1911)

- 木堆漆茶壶桶（清）
Wooden and Lacquer Tea Pot Bucket (Qing Dynasty, 1616-1911)

- 木制茶壶桶（清）
Wooden Tea Pot Bucket (Qing Dynasty, 1616-1911)

是一个小桶，内放棉絮、丝织物等保暖材料，不过桶壁上开有流口，将装满热茶的茶壶放进桶内，让壶嘴对着茶壶桶的流口，盖上盖子，在很长一段时间内可起到保温的效果。制作茶壶桶的材料多样，有竹木、藤条、丝织品等，形状或圆或方，不一而足。

茶籝

茶籝最初是一种采茶、盛茶的器具，用竹子编就。到了清代，茶籝的功用演变为装放茶器的工具，与陆羽在《茶经》中提到的都篮相当。

the Ming and Qing dynasties. It looked like a small bucket with cotton and silk inside to keep warm. However, there was an opening on the wall of the bucket. When laying the tea pot with hot tea in the bucket, you should fit the mouth of the pot to the opening of the bucket and cover the lid. Then the bucket could keep the temperature after a long while. There were various types of materials to make tea pot buckets, like bamboo, vines and silk. They could be in many shapes, like round or square.

Tea Basket

At first, the tea basket was a utensil to collect and store tea made of bamboo. In the Qing Dynasty, it was also used to store teawares, similar to teaware

- 清宫旧藏茶籝
Tea Basket in the Imperial Palace of the Qing Dynasty

- 红木茶籝（清）
Mahogany Tea Basket (Qing Dynasty, 1616-1911)

明代戏曲家高濂生性喜爱游山玩水、品茗把盏。为出游携带方便，高濂自己设计了提盒，就是茶籯。它内置茶壶、火炉、木炭，以便于随时随地品饮，平时还可以用来摆放茶具。清代时，宫廷中饮茶之风盛行，各色茶具也备受宫廷欢迎。乾隆皇帝酷爱喝茶，他一生中多次出宫南巡，为在旅途中携带方便，他特意命人制作了便于旅途用的全套茶具，并专门设计了用于装置全套茶具的茶籯（又称"撞盒"），用来放置茶壶、茶碗、茶叶罐、茶炉、水具等。故宫现存的几套茶籯，主要有纯紫檀木和竹木混制两种。这些茶籯制作精致，每一件都堪称富有创意的工艺品。

containers mentioned in Lu Yu's *The Classic of Tea*.

Gao Lian, a dramatist in the Ming Dynasty, enjoyed the sights of mountains and rivers and loved tea and teawares. In order to make drinking tea conveniently outside, Gao Lian designed a basket with handle himself. This was the tea basket. It had the tea pot, stove and charcoal inside, so people could enjoy tea at any time in any place. In ordinary days, it could be used to set teawares. In the imperial court of the Qing Dynasty, drinking tea was very popular. Thus, various teawares were welcomed by them. Emperor Qianlong was very fond of tea. He had several tours in the south. In order to make it easy to take, he asked people to make a whole set of teaware which was convenient for tours. In addition, he also asked people to design tea basket (also called "teaware box") to store all the teawares, including the tea pot, tea bowls, tea jar, stove and water wares. The tea baskets in the Imperial Palace were either made of rosewood only or bamboo together with wood. These tea baskets were delicate and each one was a very creative piece of art work.

明清绘画中的茶具
Teawares in paintings of the Ming and Qing Dynasties

在元代文人画发展的基础上，明代绘画有了继承和发展，尤其到了晚明，以"吴中四杰"文徵明、唐寅等人为代表，兴起了文人创作的高潮。文人对茶情有独钟，而茶成了文人画中不可或缺的一部分。明清两代涉及煮茶、品茶内容的画作特别多，从中我们可以一窥明清时期茶具的风貌。

Paintings in the Ming Dynasty followed and developed on the basis of paintings in the Yuan Dynasty. Especially in the late stage of the Ming Dynasty, "Four Masters from Wu" with Wen Zhengming and Tang Yin as the representatives, brought about a high tide of creation. Scholars had special interest on tea and tea became a necessary part in their paintings. There were many paintings concerning boiling tea and tasting tea in the Ming and Qing dynasties, from which we could learn teawares of that time.

明 文徵明《惠山茶会图》

文徵明（1470—1559），名璧，字徵明，号衡山居士，吴县（今江苏苏州）人。明代吴门画派代表人物之一，擅长山水、人物、花鸟画。文徵明的茶事绘画作品很多，其中《惠山茶会图》是他的代表作之一。

此画创作于明正德十三年（1518年），图中表现的是清明时节，文徵明与好友蔡羽、汤珍、王守、王宠等游于惠山（今江苏无锡西郊），在二泉亭下以茶雅集的场景。在起伏的深山中，处处是高大的松树，其间有一井亭。画面共有七人，其中四位文士，三位侍者。两位文士围井栏而坐，在井亭旁的茶桌边两侍者正在烹茶，红色漆茶桌上放着相应的茶具，一侍者正蹲踞在竹茶炉边扇火煮水，竹炉上放着一把茶壶。还有两位文士正在山中的曲径攀谈，画面充满闲适幽静淡泊的气氛。惠山泉水甘洌可口，极宜烹茶，自被唐代陆羽评为"天下第二泉"后，声名远扬。文徵明及其朋友常于此地赏景烹茶，作诗绘画。此画中流露出来的雅逸茶味，反映了明代后期文士崇尚自然清新而又不失古风的茶道格局。

The Painting of a Tea Meeting on Mountain Hui, by Wen Zhengming, Ming Dynasty
Wen Zhengming (1470-1559), also called Wen Bi, from Wuxian County (now Suzhou, Jiangsu Province) was a representative of the *Wu* Painting School in the Ming Dynasty. He was good at painting landscapes, portraits, and flower-and-bird scenery. He had many works about tea activities and *The Painting of a Tea Meeting on Mountain Hui* was one of his masterpieces.

This painting was created in the thirteenth year of the Zhengde Period of the Ming Dynasty. It described a scene that on the day of *Qingming* (Pure Brightness), Wen Zhengming had a tour on Mountain Hui (now Western countryside of Wuxi, Jiangsu Province) with his friends Lu Yu, Tang Zhen, Wang Shou and Wang Chong. They gathered together in the name of tea in a pavilion. Among the rolling mountains, there were many big pine trees and a pavilion with a well. There were seven people in this painting, four scholars and three servants. Two scholars were sitting around the fence of the well while two servants were making tea at the table next to the well. There were tea wares on the red lacquer table. One servant crouching aside the bamboo stove was fanning and boiling water. On the stove, there was a tea pot. Another two scholars were talking in the winding pathway of the mountain. This whole painting gave a relaxing and peaceful feeling. The spring water from Mountain Hui was very sweet and suitable for making tea. Since it was named "the second best spring" by Lu Yu in the Tang Dynasty, this spring became well-known. Wen Zhengming always enjoyed scenaries, made tea, composed poems and created paintings here with his friends. The relaxing feeling of tea reflected in this painting showed that scholars at the late stage of the Ming Dynasty loved a fresh yet traditional way of making tea.

- 明 文徵明《惠山茶会图》
The Painting of a Tea Meeting on Mountain Hui, by Wen Zhengming (Ming Dynasty, 1368-1644)

明 文徵明《品茶图》

此画作于明嘉靖十年（1531年）初春的谷雨节气之后。画面远处山峰耸翠、轻烟紫雾，中景为溪涧峡谷、参天古木，近处是小桥流水、茅轩草舍。茶轩中，主人与来访的得意门生陆子传（临窗而坐者）交谈，桌子上一把茶壶和两个茶碗摆在宾主面前。右侧一间小茅舍中，一童正跪在地上的风炉前煎水，准备泡茶。文徵明在画上有自题跋和小序："碧山深处绝纤埃，面面轩窗对水开。谷雨乍过茶事好，鼎汤初沸有朋来。嘉靖辛卯，山中茶事方盛，陆子传过访，遂汲泉煮而品之。真一段佳话也。徵明制。"

The Painting of Tasting Tea, by Wen Zhengming, Ming Dynasty

This painting was done after the day of *Guyu* (Grain Rain) in early spring in 1531. At the far back of this painting, there were two green mountains covered in a little fog. The medium shot was made up by a creek and a crayon, with old trees reaching to the sky. The closest part is a bridge, flowing water and thatched cottages. In the tea house, the master was talking with his student Lu Zichuan (the one sitting near the window), of whom he felt very proud. A tea pot and two tea bowls were placed in front of them on the table. In the thatched hut on the right, the boy kneeling on the ground was boiling water in front of the stove and preparing to make tea. Wen Zhengming wrote inscription and introduction on the painting: "There was no dust in the remote green mountains; all the windows were opening towards the water. It was good to drink tea after the day of *Guyu*; Friends came as the water boiled for the first time. In the tenth year of the Jiajing Period, making tea was very popular in the mountains. I took spring water to have tea with Lu Zifang who was visiting me today. This was really a good anecdote. Written by Zhengming."

• 明 文徵明《品茶图》（局部）
The Painting of Tasting Tea (Partial), by Wen Zhengming (Ming Dynasty, 1368-1644)

明 唐寅《事茗图》

唐寅（1470—1523），字伯虎，一字子畏，号六如居士、桃花庵主等，吴县（今江苏苏州）人，山水、人物、仕女、花鸟无所不工，是"吴门画派"的代表人物之一。他的系列茶画中，以《事茗图》最享盛誉，此图主要反映了明代文人的庭院书斋生活。

画面左侧有巨石山崖古木，画面正中双松之下有茅屋数间。茅屋中有一人伏案读书，案头放置一把大壶，从形制上看，应是紫砂壶。明代紫砂壶初期以大壶为主，此图正表现出这个细节。侧屋一童子正在烹茶，桌案上也放着紫砂壶以及杯、

- 明 唐寅《事茗图》

The Painting of Making Tea, by Tang Yin (Ming Dynasty, 1368-1644)

罐等茶具。舍外右方，小溪上横卧板桥，桥上有一老翁拄杖缓行，后随抱琴童子。远处群山屏列，瀑布飞流，潺潺流水绕屋而过。画面一侧题诗："日长何所事，茗碗自赍持。料得南窗下，清风满鬓丝。"表现出明代文人雅士追求远离尘俗、品茗抚琴的闲适生活的志趣。

The Painting of Making Tea, by Tang Yin, Ming Dynasty

Tang Yin (1470-1523), also named Bohu, Ziwei, and Master of Peach Blossom Hut, was from Wuxian County (now Suzhou, Jiangsu Province), He was an expert at paintings about mountains and rivers, personage, maidens, flowers and birds, also a representative of the Wu Painting School. In his tea painting series, *The Painting of Making Tea* was the most famous one. This painting mainly reflected scholars' life in their yard and study room.

On the left side of the painting, there were some huge stones, the cliff and ancient wood. In the middle part, several thatched cottages were under two pine trees. In a cottage, one person was reading over the table. The huge pot on the table should be a purple clay teapot judging from its style. At the first stage of the Ming Dynasty, purple clay teapots were always big ones, which was exactly the point of this painting. In the side hut, a boy was boiling tea with a purple clay teapot, bowls and jars on the table. At the right outside the house, a bridge crossed over the creek. On the bridge an old man was walking slowly with a stick, followed by a boy with a Chinese zither. At the far back, there were mountains and flowing waterfalls. Water went around the tea house in the creek. On the left side was a poem for this painting: "What should I do in such a long day? Having tea was a good choice. When sitting under the window on the south, I could feel breeze going through my hair." This poem showed that scholars in the Ming Dynasty pursued a peaceful and relaxed life of having tea and playing Chinese zither, leaving far from the busy world.

明 陈洪绶《品茶图》

　　陈洪绶，字章侯，号老莲，诸暨（今属浙江）人，是明代著名的人物画家。

　　面画上两位高士（一正面、一侧面）相对而坐，一坐于一片硕大的芭蕉叶上，一坐于一长方崎岖石案之后的石上。此时琴弦收罢，茗乳新沏，良朋知己，香茶在手，似评古说今已罢，正凝神片刻。正面高士左边石上有一圆肚茶壶，茶壶旁边有黑色的茶炉，里面燃着红色的炭火，茶炉上为一直柄上翘的茶壶。画面的右边露出一角莲叶和莲花。此画人物造型刚柔相济，环境幽雅宜人，把人物的隐逸情趣和文人高雅的品茶生活，渲染得既充分又得体，给人以美的享受。

The Painting of Having Tea, by Chen Hongshou, Ming Dynasty

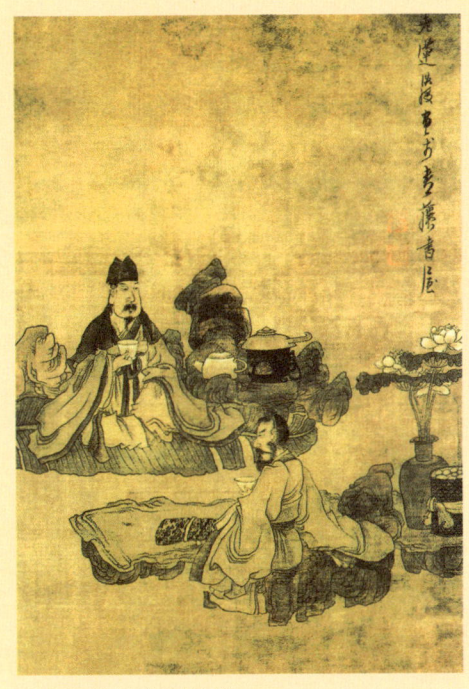

• 明 陈洪绶《品茶图》
The Painting of Having Tea, by Chen Hongshou (Ming Dynasty, 1368-1644)

Chen Hongshou, also named Zhanghou, Senior Lotus, from Zhuji (now in Zhejiang Province), was a renowned painter of personage in the Ming Dynasty.

　　In this picture, two scholars (one facing the front and the other facing a side) sat face to face. One was sitting on a huge banana leaf, the other one was sitting on a stone behind a square rough stone table. At this time, they just finished playing music while the tea was ready. With tea in hands, the two friends seemed to have just finished talking about history and present and were in deep thought. On the stone, which was on the left side of the person facing to the front, there was a round belly tea pot with a black stove aside. Red fire is burning in the stove. A tea pot with an upward handle was on the stove. On the right side of the painting, there was half a lotus flower and leaf. Characters drawn here showed both force and grace. The peaceful and pleasant environment perfectly depicted the secluded and elegant life of drinking tea while also brought an enjoyment of beauty.

清 钱慧安《烹茶洗砚图》

钱慧安（1833—1911）字吉生，祖籍浙江湖州，晚清时期著名画家，以人物画见长。

这幅画是清同治十年（1871年），作者39岁时为友人文舟所作的肖像画。在两株虬曲的松树下，有傍石而建的水榭，一中年男子倚栏而坐。榭内琴桌上置有茶具、书函，一侍童在水边洗砚，另一侍童拿着蒲扇，对小炉扇风烹茶。红泥小火炉上架着一把东坡提梁壶，炉边还放有一个色彩古雅的茶叶罐，而这时的小童正侧头观看一只飞起的仙鹤。画面的意境给人以高雅脱俗之感。

The Painting of Making Tea and Washing Inkstone, by Qian Huian, Qing Dynasty

Qian Huian (1833-911), also named Jisheng, from Huzhou, Zhejiang Province, was a famous painter in the late period of the Qing Dynasty and was good at painting personage.

In 1871, the thirty-nine-year old Qian painted a portrait for his friend Wen Zhou. There was a waterside pavilion made of stone under winding pine trees and a man in his middle age was sitting against the fence. Teawares, books and letters were set on the table inside the pavilion. One servant boy was washing the ink stone by the water while another one was making tea by fanning to the stove. On the clay stove, there was a pot with handle. A classic and beautiful tea jar was on the side of this stove. At this time, the boy was turning his head to watch a flying red-crown crane. This painting gave people an elegant and refined feeling.

- 清 钱慧安《烹茶洗砚图》
The Painting of Making Tea and Washing Inkstone, by Qian Huian (Qing Dynasty, 1616-1911)

各种材质的老茶具
Ancient Teawares of Different Materials

　　中国老茶具的制作材质非常丰富，除了最受欢迎的瓷和紫砂之外，金银、琉璃、漆、锡、景泰蓝、玉石、竹木、果壳等材质也都比较常见，这反映出中国古代茶文化与传统手工艺广泛而深入的结合。
There are various kinds of materials used in making traditional Chinese teawares. Besides the most popular porcelain and purple clay ones, gold and silver, glass, lacquer, tin, cloisonne, jade, bamboo and shell are also common in making teawares. This reflects a comprehensive and close connection between traditional Chinese tea culture and handicrafts.

> 瓷茶具

茶具这个大家族中，陶瓷茶具种类最为丰富，这同陶瓷与茶的天然契合分不开。陶瓷是中国先民的伟大发明，在距今八千年左右的新石器时代早期便出现了。伴随着人

• 青瓷茶具
Celadon Glazed Teaset

> Porcelain Teawares

In the family of teawares, porcelain has the largest number of variations. This is because of the natural link between porcelain and tea. Porcelain is a big invention of ancient Chinese people. Ceramic utensils were in use early in the first stage of Neolithic Period, about 8,000 years from now. With the development of civilization, this art of "earth and fire" continuously flourished and brought out new ideas. In the Shang Dynasty about 3,000 years ago, original porcelain emerged. Then after thousands years of creation and accumulation, in the Eastern Han Dynasty, craftsmen first made mature celadon in Shangyu, Zhejiang Province. The model of this celadon used kaolin clay and porcelain stone and was fried in high temperature of 1200℃ –1300℃. Thus the body was hard, tight and water-resistant. There

类文明的进步，这一"土与火"的艺术不断丰富、成熟、推陈出新。到三千多年前的商代，出现了原始瓷器。又经过千余年的创造积累，到东汉时期，浙江上虞一带的工匠首先创烧出成熟的青瓷器。这种青瓷器的坯体采用了高岭土和瓷石等材料，在1200℃～1300℃的高温中烧制而成，胎体坚硬致密不吸水，胎体外面罩施一层釉，釉面光洁，呈青灰色。自东汉后期成熟的瓷器产生后，瓷器就以其耐高温、生产量大、价廉、洁净等特点成为大众的生活用品。而茶性洁的特点与瓷器的洁净很相合，所以瓷器一经产生就同茶联系在一起。

青瓷茶具

青瓷是一种高温颜色釉瓷器，以铁为呈色剂。青瓷茶具因色泽青翠，用来冲泡绿茶，更有益汤色之美，所以一直以来广受欢迎。

唐代时，越窑生产的青瓷釉质如玉、釉色青绿闪黄，在唐代风靡一时，被视为"瓷器中的贵族"。

宋代，作为当时五大名窑之一的浙江龙泉窑生产的青瓷茶具，已

was also a layer of green and grey glaze outside, which was smooth and bright. Since a mature production of porcelain in the late stage of the Eastern Han Dynasty, porcelain became a daily utensil for common people, heat-resistant, mass produced, affordable and easy to clean. In addition, both tea and porcelain shared a characteristic of cleanness. Therefore, porcelain was naturally linked to tea when it first appeared.

Celadon Porcelain Teasets

Celadon is a high-temperature color glazed porcelain with iron as its coloration. Since celadon teaware looks fresh and green and contributes to the look of tea, it is always welcomed by people in different times.

- 越窑青釉划花托盏（北宋）
Celadon Glazed Tea Bowl and Base from *Yue* Kiln (Northern Song Dynasty, 960-1127)

然达到鼎盛时期，远销各地。龙泉窑的工匠们烧制出粉青、梅子青这两种代表性的精品。粉青、梅子青釉层中含有大量小气泡和未完全熔化的石英颗粒，当光线射入釉层时，釉面会呈现出温润如玉、青翠如梅的色泽，从而使龙泉青瓷达到极高的艺术境界。南宋末期，龙泉窑曾一度因战乱而停烧。但到了元代，龙泉青瓷的生产不仅得以恢复，而且蒙古族粗犷豪放的风格也融入到瓷器的制造之中，器物胎质浑厚，釉色成熟凝重。明代龙泉

- 龙泉窑青釉刻花莲花提梁壶（明）
  Celadon Glazed Tea Pot with Ridge and Engraved Lotus Pattern from *Longquan* Kiln (Ming Dynasty, 1368-1644)

In the Tang Dynasty, the glaze of celadon from *Yue* Kiln looked similar to jade, which had slight yellow within dark green. Celadon then was fashionable for a time and considered an "aristocrat in porcelain".

In the Song Dynasty, as one of the five famous kilns at that time, *Longquan* Kiln in Zhejiang Province produced celadon tea sets which reached a peak and sold throughout the country. Craftsmen from *Longquan* Kiln made two selected porcelain, pale blue and plum green. Pale blue and plum green glaze had a large sum of bubble and unmelting quartz particles. When light shone on the glaze, it would show a color soft as jade and green like plum. This effect helped celadon from *Longquan* Kiln reach an extremely high position in art. At the late stage of the Southern Song Dynasty, *Longquan* Kiln stopped its production for a while. However, in the Yuan Dynasty, not only the production of celadon from *Longquan* Kiln was recovered, but also the unconstrained and heroic style of Mongolian people was also integrated into the porcelain making. The base of this celadon was very deep and thick and the color of the glaze became mature and dignified. In the Ming Dynasty, celadon

瓷厚实雄浑中不乏秀美典雅，在烧造工艺上达到了一个新的高度。16世纪末，龙泉青瓷出口法国，轰动整个法兰西，人们用当时风靡欧洲的名剧《牧羊女》中的女主角雪拉同的美丽青袍与之相比，称龙泉青瓷为"雪拉同"，视其为稀世珍品。

from *Longquan* Kiln shared both the features of vigor and elegance, which came to a new level in art skills. In the late sixteenth century, celadon from *Longquan* Kiln was exported to France and made a big hit there. At that time, French people compared it to Celadon, the heroine of a French opera, and took it as a rare treasure.

- 龙泉窑青釉莲瓣纹盖杯（南宋）
Celadon Glazed Cup and Cover with Lotus Leaf Pattern from *Longquan* Kiln (Southern Song Dynasty, 1127-1279)

白瓷茶具

白瓷，诞生于三国两晋南北朝时期，在北朝晚期，北方的邢窑利用当地特有的瓷土烧制出洁白细腻的白瓷，打破了青瓷独统天下的局面。白瓷是由青瓷演变而来，所以最初的白瓷茶具胎呈浅黄褐色，釉呈乳白色泛青黄，积釉处为青色，釉层薄而滋润。自唐以来生产白瓷

White Porcelain Teawares

White porcelain emerged during the period of the Three Kingdoms Period, the Jin and the Southern and Northern dynasties. In the late Northern dynasties, *Xing* Kiln in the north used their unique porcelain clay to make white and fine white porcelain, which invaded the world of celadon. White porcelain evolved from celadon. Early white porcelain

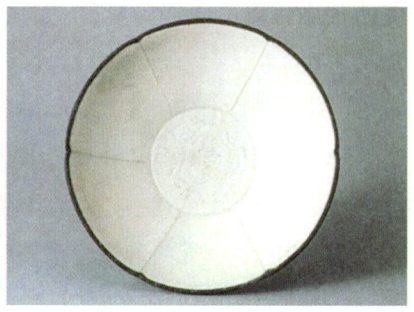

- 定窑白釉刻花碗（宋）
White Glazed Bowl with Engraved Pattern from *Ding* Kiln (Song Dynasty, 960-1279)

- 德化窑白釉小盏（明）
White Glazed Tea Bowl from *Dehua* Kiln (Ming Dynasty 1368-1644)

茶具的窑场很多，如河北任丘的邢窑、浙江余姚的越窑、湖南的长沙窑、四川的大邑窑等，但最为著名的当属江西景德镇出产的白瓷茶具，它以"白如雪、薄如纸、明如镜、声如磬"而闻名于世。

teawares were light yellowish brown with milky white and bluish yellow glaze. The places where glaze gathered show blue and where glaze was thin and looked moist. Since the Tang Dynasty, there were many kilns which made white porcelain teasets, for example, *Xing* Kiln in Renqiu, Hebei Province, *Yue* Kiln in Yuyao, Zhejiang Province, *Changsha* Kiln in Hunan Province and *Dayi* Kiln in Sichuan Province. However, the most famous white porcelain teawares were made by *Jingdezhen* Kiln in Jiangxi Province. They were world renowned for the features of their porcelain, which was "as white as snow, as thin as paper, as bright as mirror and as loud as Chime stone".

黑瓷茶具

　　黑瓷茶具，始于晚唐，鼎盛于宋，延续于元，衰微于明、清，这是因为自宋代开始，饮茶方法由唐时的煎茶法逐渐改变为点茶法，而宋代流行的斗茶，又为黑瓷茶具的崛起创造了条件。宋时福建的建窑、江西的吉州窑、山西的榆次

- **建窑曜变茶盏（宋）**
  曜变茶盏是建窑黑釉茶器中的珍贵品种，其外形端庄，盏内外壁黑釉上散布浓淡不一、大小不等的琉璃色斑点，光照之下，釉斑会折射出晕状光斑，十分神奇。曜变的形成原理很特殊，在烧窑中釉水发生变化，这种变化系偶然出现，非窑工人力可为，因此成品极为罕见。

  Tea Bowls with Altered Glaze from *Jian* Kiln (Song Dynasty, 960-1279)
  The tea bowl with altered glaze is a precious one among black-glazed teawares from *Jian* Kiln. It looks dignified and has colored glaze spot of different shades and sizes everywhere on the black glaze. Glazed spot refracts amazing halo light spot. The theory of forming the altered glaze is very special. Glaze water changes in the kiln and this change happens by accident. People can not control the process, thus it is rare to see the products.

Black Porcelain Teawares

Black porcelain teawares began in the late Tang Dynasty, reached its peak in the Song Dynasty, continued in the Yuan Dynasty and finally declined in the Ming and Qing dynasties. The reason was that the way of having tea gradually changed from cooking tea to tea dripping since the Song Dynasty. In addition, the prevalence of tea contest in the Song Dynasty created positive condition for the rising of black porcelain teaware. In the Song Dynasty, many kilns like *Jian* Kiln in Fujian Province, *Jizhou* Kiln in Jiangxi Province and *Yuci* Kiln in Shanxi Province, all produced a large number of black porcelain teawares and became

- **建窑黑釉兔毫盏（宋）**
  兔毫因茶盏内外釉面结晶而现出细长如兔毛状的纹路而得名，又根据兔毫盏色泽的微妙不同分"金兔毫""银兔毫"和"黄兔毫"。

  Black-glazed Bowl with Hare Fur Pattern from *Jian* Kiln (Song Dynasty, 960-1279)
  Hare fur porcelain is named because the patterns on the glaze look like thin and long hare fur. Because of the tiny differences in color, hare fur bowls are further distinguished into "golden hare fur" "silver hare fur" and "yellow hare fur".

窑等窑口，都大量生产黑瓷茶具，成为黑瓷茶具的主要产地。黑瓷茶具的窑场中，建窑生产的黑瓷茶盏最为人称道。建窑茶盏俗称为"黑建"或"乌泥建"，其胎质以灰黑色为主，釉层厚，并有流釉现象，由于釉面配料的缘故而呈结晶状，呈色变化万千，其中最有代表性的是兔毫、油滴、曜变、鹧鸪等品种。

the main source of it. Among all the kilns that produced black porcelain tearwares, *Jian* Kiln won most praise. Tea bowls from *Jian* Kiln was also called "black bowls" or "mud bowls". Black was the main color of the porcelain base. It had thick glaze and also a "flowing glaze" phenomenon. Because of the ingredients in the glaze, the crystal shape glaze could show thousands of different forms, especially hare fur, oil drop, altered glaze and partridge.

- 黑釉油滴盏（南宋）

油滴是建窑黑釉茶器中的名贵珍品，油滴盏的釉面密布着呈银灰色金属光泽的小圆点儿，因形似油滴而得名。其形成原因很复杂，因此存世量不多。油滴的形成是由于铁的氧化物高温下在釉面富集，冷却后以赤铁矿和磁铁矿的形式从中析出晶体所致。油滴盏如黑夜星辰，闪烁变幻，受茶人、收藏家垂青。

Black-glazed Bowl with Oil Drop Pattern (Southern Song Dynasty, 1127-1279)

Oil drop porcelain is the treasure in black-glazed teawares. Silver grey spots spread over the glaze of the bowl. The bowl gets the name as the shape of the spot looks like oil drop. Since it is very complicated to get this glaze, there are only a few left at the moment. The shape of the oil drop comes from the gathering of iron oxide on glaze under high temperature, which separates out crystals in forms of hematite and magnetite after cooling. The oil drop bowl looks like a starry night, shinning and dazzling, much appreciated by tea loers and collectors.

- 鹧鸪斑黑釉盏（宋）

鹧鸪斑也是建窑黑釉盏中的极品，在黑色釉面上分布大小不均的白色圆形斑点；黑白分明，视觉冲击力十分强烈。其形成原因有两种，一种是自然生成，属于窑变类，另一种则是以白釉点染而成。物以稀为贵，自然形成的贵重些。

Black-glazed Bowl with Partridge Pattern (Song Dynasty, 960-1279)

The partridge pattern is also a masterpiece in black-glazed bowls from *Jian* Kiln. White spots of different shapes spreading over the black glaze have a great impact to sight, which forms a sharp contrast between black and white. There are two reasons for its formation. One is naturally generated, which belongs to the change of glaze color in the kiln, and the other is dyed with white glaze. Rore things are precious, especially those natural forms.

其他单色釉瓷茶具

中国的瓷质茶具，可以分为单色釉瓷茶具和彩瓷茶具两大类。而单色釉瓷茶具中，除了青釉、白釉和黑釉三个比较多见的品种之外，还有青白釉、红釉、黄釉、蓝釉、绿釉等其他颜色的种类。

Other Single Color Glazed Porcelain Teawares

Chinese porcelain teawares can be divided into single-color-glazed and multicolor glazed ones. Besides celadon, white and black glaze, which are the most common single colors, there are also other types like green and white, red, yellow, blue and green glazes.

- 青白釉刻花双鱼纹碗（南宋）

   Bluish-white Glazed Bowl with Engraved Double-fish Pattern (Southern Song Dynasty, 1127-1279)

- 青白瓷执壶（宋）

   青白釉瓷创烧于宋代的景德镇，也叫"影青"，釉色介于青白二色之间，青中泛白、白中透青。

   Bluish-white Glazed Pot with Handle (Song Dynasty, 960-1279)

   Bluish-white glaze porcelain, also called "Yingqing," was created in Jingdezhen in the Song Dynasty. The color of this glaze is between blue and white, each shade staying within the other one.

- 珊瑚红釉盏（清 雍正）

   珊瑚红是一种以铁为着色剂的低温红釉，因其呈色红中闪黄，与珊瑚颜色相似，故而得名。

   Coral-red-glazed Tea Bowl (Yongzheng Period of the Qing Dynasty)

   Coral red is a low-temperature red glaze with iron as its colorant. It gets the name because this glaze has slight yellow in red, which is similar to the color of coral.

- **霁红釉小碗（清 乾隆）**

  霁红是以铜为着色剂的高温红釉，创烧于清康熙朝后期，成品色调深红，似暴雨后晴空中的红霞，故名"霁红"。其特点是釉汁凝厚，釉面润泽，有桔皮纹。

  Sunglow-red Glazed Bowl (Qianlong Period of the Qing Dynasty)

  Shiny red glaze is a high-temperature red one with copper as its colorant. It was created at the late stage when Emperor Kangxi was in power. The glaze gets this name because the finished product shows dark red, which looks like the red sky after a storm. It has the feature of thick glaze liquid, smooth glaze and orange peel pattern.

- **黄釉杯及杯托（清）**

  黄釉是一种以铁为着色剂的低温釉。在清代，纯正的黄色釉瓷器是皇家专用的。

  Yellow-glazed Cup and Base (Qing Dynasty, 1616-1911)

  Yellow glaze is a low-temperature glaze with iron as its colorant. In the Qing Dynasty, pure yellow-glazed porcelain could only be used by royal family members.

- **黄地绿团龙纹碗（清）**

  Yellow-glazed Bowl with Green Loong in a Circle Pattern (Qing Dynasty, 1616-1911)

- **绿釉菊瓣形盖碗（清 乾隆）**

  绿釉是含氧化铜的石灰釉，在还原气氛中呈红色，在氧化气氛中则呈绿色。我国传统的绿釉和绿彩都是以铜作为着色剂，属于铜绿釉。

  Green-glazed Bowl and Cover with Chrysanthemum Leaves Pattern (Qianlong Period of the Qing Dynasty)

  Green glaze is lime glaze with cupric oxide. It looks red in the process of deoxidization and green in the process of oxidation. Traditional Chinese green glaze and green color both use copper as their colorant.

彩瓷茶具

彩瓷，又称为"彩绘瓷"，是在器物表面加以彩绘的瓷器，主要有釉下彩瓷和釉上彩瓷两大类。釉下彩瓷，是指彩色纹饰呈现在瓷器表面透明釉的下面，其特点是画面不暴露于外界，既不会被磨损和腐蚀，又不致有污染的危害。釉下彩瓷中最著名的要算成熟于元代的青花瓷、釉里红瓷、青花釉里红瓷，还有清代创制的釉下三彩、釉下五彩等。而釉上彩瓷的彩色纹饰呈现在瓷器表面的釉面之上，在装饰上由简单到复杂，色彩由一种到多种，鲜艳光亮，装饰性更强。主要的釉上彩瓷品种有斗彩、釉上五彩、粉彩、珐琅彩等。

青花瓷茶具

在品种繁多的彩瓷茶具中，尤以青花瓷茶具最引人注目。青花瓷，是指以氧化钴为呈色剂，在瓷胎上直接描绘图案纹饰，再涂上一层透明釉，然后在窑内经1300℃左右高温还原烧制而成的器具。它的特点是：花纹蓝白相映成趣，色彩淡雅可人，令人感到赏心悦目。由

Faience Porcelain Teaware

Faience porcelain, also called "colored-painted porcelain", refers to the porcelains with colored paintings, including underglaze faience porcelain and overglazed porcelain. Underglaze faience porcelains mean the colored pattern shown under the transparent glaze. Thus, the colored pattern will not expose to the outside and it will neither be worn, corroded nor corrupted. Among underglaze faience porcelains, the most famous kinds are blue-and-white porcelains, underglazed red porcelain, blue-and-white underglaze red porcelains, which became mature in the Yuan Dynasty, as well as three-colored underglazed porcelains and colorful underglazed porcelains in the Qing Dynasty. The colored pattern of overglaze faience porcelains shows on the glaze. The decoration patterns vary from simple to complicated while there are bright single or multiple colors, which is more suitable for decoration. Overglazed porcelains include clashing color glazes, overglaze polychrome, famille rose and enamel glazes.

于是在青花之上涂釉，显得滋润明亮，更增添了青花茶具的魅力。

　　青花瓷诞生于元代以前，但直到元代中后期，青花瓷茶具才开始成批生产，特别是江西景德镇，成了中国青花瓷的主要生产地。由于青花瓷绘画工艺水平高，特别是将

Blue-and-white Porcelain Teasets

Among various kinds of faience tea wares, blue-and-white porcelains attract the most attention. They refer to the patterns made directly on the porcelain bases and added a layer of transparent glaze with oxidized cobalt as colorant. They will be deoxidized under 1300℃ and then the porcelain will be completed. Blue-and-white patterns set each other off, simple but elegant. Since the pattern is made over the glaze, it looks brighter, which also adds charm to blue-and-white porcelain tea wares.

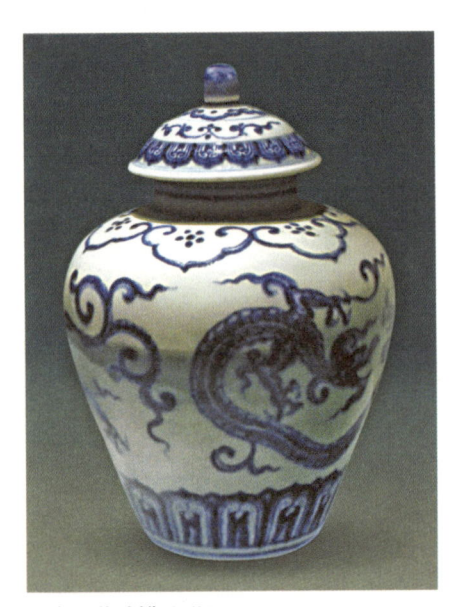

- 青花花鸟纹茶杯（明 万历）
Blue-and-white Porcelain Teacups with Flower and Bird Pattern (Wanli Period of the Ming Dynasty)

- 青花茶叶罐（明）
Blue-and-white Porcelain Tea Jar (Ming Dynasty, 1368-1644)

中国传统绘画技法运用在瓷器上，因此这也可以说是元代绘画的一大成就。明代，景德镇生产的青花瓷茶具，诸如茶壶、茶盅、茶盏等，花色品种越来越多，质量愈来愈精，无论是器形、造型、纹饰等都冠绝全国，成为其他青花瓷窑场模仿的对象。清代，特别是康熙、雍正、乾隆时期，青花瓷茶具在古陶瓷发展史上，又达到了一个历史高峰。

Blue-and-white porcelain was created before the Yuan Dynasty. However, not until the mid and late Yuan Dynasty, did people begin to make them in mass production, especially Jingdezhen, Jiangxi Province, which became the main production area of blue-and-white porcelain in China. Since the painting skill of blue-and-white porcelain was at the very high level, especially its application of traditional Chinese painting skills to porcelains, it could be taken as great achievements of painting skills in the Yuan Dynasty. Then in the Ming Dynasty, blue-and-white porcelains produced in Jingdezhen, like tea pots, teacups without handles and tea bowls had more kinds of patterns and higher quality. In terms of shape, model or pattern, blue-and-white porcelain was the best throughout the country and the model for other kilns which also made this kind. In the Qing Dynasty, especially during the reign of Emperor Kangxi, Yongzheng and Qianlong, blue-and-white porcelain teasets reached another peak in the history of ancient porcelain.

- 青花花鸟纹茶壶（明 万历）
Blue-and-white Porcelain Teapot with Flower and Bird Pattern (Wanli Period of the Ming Dynasty)

玲珑瓷

玲珑瓷创烧于明宣德年间（1426—1435），是在镂空工艺的基础上创造和发展起来的。在细薄的坯胎上，按照设计的图案花形，镂雕成许多米粒状的孔，然后施以透明釉，再通体施釉，入窑烧制而成。烧成后的孔眼明澈透亮，十分美观。清代瓷工们把青花瓷和玲珑瓷巧妙地结合起来，烧制出独一无二的青花玲珑瓷。碧绿透明的玲珑和古朴翠蓝的青花互相映衬，精美绝伦。

Linglong Porcelain

Linglong porcelain, created during the Xuande Period of the the Ming Dynasty (1426-1435), was a porcelain invented and developed on the basis of engraving. People engraved many rice-size holes on the thin porcelain base and added transparent glaze, then added glaze on the whole body and fired in the kiln. The completed porcelain had transparent and bright holes, which looked very beautiful. Craftsmen in the Qing Dynasty skillfully combined it with blue-and-white porcelain and made the unique blue-and-white *Linglong* porcelain. The green transparent *Linglong* and the ancient unsophisticated blue-and-white flowers set each other off, which was unmatched by others.

- 青花玲珑瓷茶具
  Blue-and-white *Linglong* Porcelain Teaset

釉里红茶具

釉里红又名"釉下红"，烧成于元代景德镇窑，是釉下彩的著名品种之一。它是用铜红料作为呈色剂，在白胎上直接绘制各种图案纹饰，施透明釉，在高温中一次烧

Underglaze Red Porcelain Teaware

Underglaze red, also called "red below the glaze", was first completed in *Jingdezhen* Kiln in the Yuan Dynasty and was one of the most famous underglazed porcelains. People painted different patterns directly on the white porcelain base with copper red as the colorant. Then they added transparent glaze and fired it in high temperature. Underglaze red pattern could be used by itself or together with blue-and-white patterns.

Blue-and-white underglaze red procelain, which was also called "blue-and-white pattern with purple porcelain", was made by painting both blue, white

- 景德镇窑釉里红牡丹纹碗（明）
Jingdezhen Kiln Underglaze Red Porcelain Bowl with Peony Pattern (Ming Dynasty, 1368-1644)

- 青花釉里红茶叶罐（清）
Blue-and-white Underglaze Red Porcelain Tea Jar (Qing Dynasty, 1616-1911)

- 青花釉里红四桃茶壶（清 康熙）
Blue-and-white Underglaze Red Tea Pot with Four-peaches Model (Kangxi Period of the Qing Dynasty)

成，既可单独装饰，也可与青花结合使用。

青花釉里红俗称"青花加紫"，是用青花和釉里红两种釉下彩在生坯上绘制纹样，罩透明釉后入窑，在高温中一次烧成。画面由红、蓝两色组成，既有青花素雅的特色，又有釉里红瑰丽的风味，青红相间、冷暖相衬、动静相映。

五彩茶具

五彩是釉上彩绘方法的一种，以红、黄、绿、蓝、紫等各种彩料按图案纹饰需要绘制在釉上，再在炉中二次焙烧而成。"五彩"不一定是五种色彩皆备，但画面中红、黄、蓝三色必不可少。明代嘉靖、万历朝的五彩笔法洒脱粗放、色彩

glaze and underglaze red glaze on the base. After adding transparent glaze, it would be completed when fired once in high temperature. The pattern was mixed by red and blue. Therefore, it had the simple and elegant feature of blue-and-white pattern as well as beautiful and implicative feature of underglaze red. Blue and red went together, a contrast of cold and warm colors.

Polychrome Porcelain Teaware

Polychrome is a method of overglaze. Craftsmen paint colorful patterns in red, yellow, green, blue or purple on the glaze and complete it by firing it in the kiln for a second time. Polychrome porcelain does not necessarily mean it has all five colors, yet red, yellow and blue are required. In the Jiajing and Wanli periods of the

- **乌金釉开光五彩安居纹壶（清 雍正）**
  这件茶壶在乌金釉地上留有树叶形开光，开光内绘五彩鹌鹑和菊花，鹌鹑取"安"音，菊花取"居"音，谐音"安居"。
  Frame Outlining Golden Black Polychrome Glazed Pot with *Anju* (Partridge and Chrysanthemum) Pattern (Yongzheng Period of the Qing Dynasty)
  This tea pot has a leaf-shape pattern on the golden black glaze. Within the pattern of chrysanthemum and a colorful partridge, it is named *Anju* (live peacefully) as homonym of the two things: *An* (peacefully) from *Anchun* (partridge) and *Ju* (live) from *Juhua* (chrysanthemum).

艳丽，被称为历代五彩之冠。当时五彩并不是纯粹的釉上彩瓷，其中蓝色要通过釉下青花表现，所以又称为"青花五彩"。清康熙时，景德镇的制瓷匠师用釉上蓝彩代替了青花，烧制出真正的釉上五彩，还大胆地把黑色也运用到五彩上来。当时，五彩器上所绘人物的发鬓都用黑色，线条工整，形象更加生动精细。

Ming Dynasty, colorful overglaze skill was free, extensive, bright and beautiful, which ranked the first one among colorful porcelain in history. At that time, colorful glaze was not absolute overglaze because "blue" needed showing by blue-and-white underglaze. Thus it was also called "blue-and-white colorful porcelain". Then in the Kangxi Period of the Qing Dynasty, craftsmen in *Jingdezhen* Kiln replaced blue-and-white glaze with blue overglaze and made the real colorful overglaze porcelain. They also bravely applied black to colorful porcelain. At that time, the hair of figures on the colorful porcelains was black, looking neat and orderly, vivid and fine.

- **五彩十二花神杯（清）**
十二月花神杯共有12只，分别以水仙、玉兰、桃花、牡丹、石榴、荷莲、兰草、桂花、菊花、芙蓉、月季和梅花为主题，代表一年十二个月，并配以相应的诗文装饰，历来被视为康熙时期官窑瓷器中的名品。

Twelve Polychrome Porcelain Cups with Flora Pattern (Qing Dynasty, 1616-1911)
This tea set consists of twelve tea cups with different flora, including narcissus, magnolia, peach blossom, pomegranate, water lily, orchid, osmanthus fragrance, chrysanthemum, lotus, Chinese rose and plum blossom, which stand for the twelve months in a year respectively. People also compose corresponding poems to decorate. Therefore, they are always taken as a masterpiece among porcelains from Official Kilns in the Regin of Emperor Kangxi of the Qing Dynasty.

开光装饰

"开光"是中国瓷器常用的装饰方法之一。为了突出器物上的某一装饰形象，往往在器物的某一部位画出某一形状(如扇形、蕉叶形、菱形、心形、桃形、方形、圆形等)的边框，然后在该边框的空间里饰以花纹，这种技法称为"开光"，具有突出主体、对比强烈、以静衬动等特点。

Frame Outlining Decoration

Frame outlining is a skill often used to decorate porcelains. In order to stress one pattern of the porcelain, people draw the frame of this pattern(like the shape of a fan, a banana leaf, diamond, heart, peach, square or round) on one part of the utensil and decorate pattern within this frame. This skill is called frame outlining, which can outstand the theme, contrast and highlight action with still.

斗彩茶具

斗彩是釉下青花与釉上彩相结合的彩绘瓷工艺。斗彩茶具胎质洁白细腻，体量小巧薄轻，白釉柔和莹润，色彩绚丽丰富，从明代起就是非常名贵的品种，一直受到皇室贵族和文人士大夫的喜爱。明朝成化时期，景德镇工匠尝试用青花在白色瓷胎上勾勒出图案的轮廓线，罩透明釉高温烧制后，再在青花轮廓内填充色料，再次入炉低温烧成。色料有艳如血的鲜红、色嫩而闪绿的鹅黄、色深而闪青的松绿、色浓而无光的姹紫等，十分丰富。

Contending-color Porcelain Teaware

Contending-color porcelain uses color painting skill to combine both blue underglaze and white overglaze. The bodies of such teaware are white and fine, small and light. White glaze looks soft and bright while other parts are colorful. Since the Ming Dynasty, clashing color porcelain has always been precious treasures, which have been deeply loved by royal people and scholars. Ever from the Chenghua Period of the Ming Dynasty, craftmen from *Jingdezhen* Kiln tried to draw the frame of pattern with blue-and-white color on

• 斗彩蟠桃提梁壶（清 雍正）
Contending-color Porcelain Pot with Handle and Flat Peach Pattern (Yongzheng Period of the Qing Dynasty)

• 斗彩岁寒三友壶（清）
Condtending-color Porcelain Tea Pot with "Three Friends in the Winter (Pine, Bamboo and Plum)" Pattern (Qing Dynasty, 1616-1911)

清代在康熙、雍正、乾隆三朝又迎来一个斗彩的高峰。此时的斗彩色彩齐备，色与色之间的组合和谐，色彩本身便可表现出瓷器装饰的主题，达到了出神入化的境界。

粉彩茶具

粉彩，是在五彩的基础上，结合中国画的技法发展而成的低温釉上彩。从施彩的方法来看，是先在素胎上勾出图案的轮廓，在轮廓内填上一层叫做"玻璃白"的色料，再在玻璃白上堆填各种色料，用干

white porcelain body. Then, they added transparent glaze and fired it in high temperature. After that, they would draw other colors within the blue-and-white frame and put the base to fire in low temperature. There were various colors among clashing color porcelains, like red as bright as blood, shining and green goose yellow, dark and shining pine green or dense and dull purple. During the reign of Emperor Kangxi, Yongzheng and Qianlong, clashing color porcelain witnessed another peak in its history. At this moment, this porcelain had all kinds

净的笔轻轻地将颜色洗染成深浅不同的层次。彩绘颜色粉润柔和，画面细腻工整，形象生动逼真，并有浮雕之感。粉彩茶具在清康熙年间

of colors. Craftsmen combined different colors harmoniously. Therefore, colors themselves could express the theme of the porcelain, which was extremely miraculous.

Famille Rose Porcelain Teaware

Famille Rose porcelain is an overglaze porcelain developed on the base of colorful porcelain with Chinese painting. In term of glazing, people need to draw the frame of the pattern on the base and fill it with a "glass white" color. Then they add other colors on the glass white. The next step is to wash and dye the different colors with a clean brush into various layers. Famille rose porcelains are featured with soft and delicate color,

- 粉彩花鸟纹盖碗（清）
Famille Rose Porcelain Bowl and Cover with Bird Pattern (Qing Dynasty, 1616-1911)

- 粉彩莲花盖碗（清）
Famille Rose Porcelain Bowl and Cover with Lotus Pattern (Qing Dynasty, 1616-1911)

- 紫金釉开光粉彩花鸟纹壶（清 乾隆）
Purple Golden Frame Outlining Famille Rose Porcelain Tea Pot with Flower and Bird Pattern (Qianlong Period of the Qing Dynasty)

诞生后，立即受到宫廷和民间的广泛喜爱，成为当时高档茶具的代表类型。

珐琅彩茶具

珐琅彩又称"瓷胎画珐琅"，是清康熙时期的工匠们移植铜胎画珐琅的工艺烧制而成的彩瓷品种。珐琅彩瓷的制作不同于其他瓷器，

orderly drawings, vivid images and an anaglyph effect. It was created during the Kangxi Period of the Qing Dynasty. Soon after, it was warmly welcomed by the royal family and common people and became the model of top grade teawares.

Enamel-color Porcelain Teaware

Enamel-color porcelain, also called "enamel on the porcelain base", is one kind of faience porcelain created in the Kangxi Period of the of the Qing Dynasty. At that time, craftsmen applied the skill of enamel on copper base to porcelain and invented this technique. Enamel-color porcelain was different from other porcelains. Craftsmen in *Jingdezhen* Kiln were only responsible for making the porcelain bases. Paintings would be done by royal painters in Beijing. At last, enamel workshop belonging to imperial palace would fire them for a second time. Enamel-color porcelain had white base, soft glaze, bright color and the technique to make it surpass other porcelains. As precious porcelain for royal families, it was ruled

- 青花珐琅彩花卉纹茶叶罐（清）
Enamel-color Blue-and-white Tea Pot with Flower Pattern (Qing Dynasty, 1616-1911)

- 珐琅彩茶壶（清 雍正）
Enamel-color Porcelain Tea Pot (Yongzheng Period of the Qing Dynasty)

景德镇瓷工仅是制作素胎，绘画由远在北京皇宫的宫廷画师完成，最后由清宫造办处的珐琅作坊进行二次烧制。珐琅彩瓷胎白釉润，色调明快，其技艺精湛程度远非其他瓷器所能比。作为名贵的御用瓷器，清代规定珐琅彩瓷仅供皇帝、后妃玩赏，以及祝寿、祭祀之用。然而这一精美瓷器的烧制时间却十分短暂，只经历了康熙、雍正、乾隆三朝，乾隆以后，这种工艺便失传了。

广彩茶具

"广彩"是广州地区釉上彩瓷的简称，始于清康熙年间，成熟于乾隆时期，300多年来，广彩一直是中国外销瓷的主要品类之一，备

that in the Qing Dynasty, enamel-color porcelains could only be held by the emperor and his wives, or for the purpose of offering birthday congratulation or sacrifices to gods or goddesses. However, this delicate porcelain only had a very short life. After the Kangxi, Yongzheng and Qianlong periods, this craft was lost forever.

- **紫砂珐琅彩茶壶（清 康熙）**
  Purple Clay Enamel-color Porcelain Tea Pot (Kangxi Period of the Qing Dynasty)

- **广彩满大人庭园消闲图壶（清 乾隆）**
  "满大人"一词出现于17世纪晚期，原是西方人对中国清代各级官员的称呼。清代外销瓷茶壶上出现清装人物图案后，西方人把这种清装人物纹也称为"满大人"。
  Guang-color Porcelain Tea Pot with Mandarins Relaxing in the Yard Pattern (Qianlong Period of the Qing Dynasty)
  The word "*mandarin*" was first in use in the late 17th century and used to be a form of address by westerners to call Qing-dynasty officials. When exported porcelain pots had personage in Qing-dynasty dress patterns, westerners also called these figures "*mandarin*".

受国外收藏者追捧，在欧美大型博物馆里几乎都能看到广彩瓷器的藏品。广彩瓷运用织锦图案的手法，利用各种颜色和金银水进行钩、描、织、填，显得构图紧密、色彩浓艳、金碧辉煌，犹如万缕金银丝织在白玉上，所以又被称为"织金彩瓷"。

Guang-color Porcelain Teaware

"*Guang*-color porcelain", short for Guangzhou overglazed porcelains, started from the Kangxi Period and was mature in the Qianlong Period. In the past three hundred years, *Guang*-color porcelain has always been one of the main exported porcelains. Warmly pursued by foreign collectors, it can be seen in almost every European and American museums. *Guang*-color porcelain uses the skill of brocade pattern, which applies various colors, gold and silver water to crochet, retouch, knit and fill. Therefore, the composition of the picture looks intense, bright, colorful and splendid, like thousands of gold and silver silk woven on the white jade. So it is also called "gold weaving colorful porcelain".

- 广彩洋人狩猎图壶（清 乾隆）

广彩茶具的装饰题材除中国传统的花鸟虫鱼、山水人物外，为了适应外销的需求，也常见西方题材的故事、风景、人物等，还有部分完全按照外商提供的图样生产。

Guang-color Porcelain Pot with Westerners Hunting Pattern (Qianlong Period of the Qing Dynasty)

Besides traditional Chinese flower, bird, insect, fish, mountain, water and personage, in order to meet the need of export, decoration patterns on *Guang* color porcelains also include western stories, scenery and personage. In addition, part of the porcelains are made according to the model provided by foreign customers.

墨彩茶具

墨彩也是釉上彩的一种，是用黑色彩料在瓷胎釉面上绘画，再入炉烘烤，色料经久不脱落。墨彩茶具始见于清代康熙朝中期，流行于雍正、乾隆朝并一直延续至清末。康熙时的墨彩色泽浓重，彩釉配制纯净，烧成后漆黑莹亮。墨彩茶具的纹饰多以花鸟为主，画风深受同时代画家的影响，犹如白纸作画，浓淡相宜，洁净素雅。

Black-color Porcelain Teaware

Black-color, one kind of overglazed porcelains, refers to porcelains that use black color to paint on the glaze of the base and then are fired in the stove. The color will stand for a very long time. Black color porcelain teaware started in the mid stage of the Kangxi Period of the Qing Dynasty, became popular in the Yongzheng and Qianlong periods, and this trend continued till the end of the Qing Dynasty. In the Kangxi Period, black color was dark while the color glaze was pure. Thus the porcelains looked black and bright when finished. Black color porcelain tewares mainly used flower and bird patterns, which was deeply influenced by the painters at that time. These patterns look like writings on white paper, which show a perfect mix of dark and light, looking clean and elegant.

- 墨彩山水书法茶壶（清）
  Black-color Porcelain Tea Pot with Mountain and River Pattern and Calligraphy (Qing Dynasty, 1616-1911)

老北京的大碗茶

　　大碗茶风行于20世纪30—40年代的老北京，源于旧时民间的野茶摊。这是用大壶或大桶冲泡的茉莉花茶，由卖茶人用水舀子舀在粗瓷的蓝边大碗里，盖上玻璃片待售。路过的人如果口渴了，扔给卖茶人几分钱，端起碗来猛灌一气，解渴消暑，神清气爽。这种清茶不讲究好茶好水，只为解渴，颇有一番粗犷的民间野趣。卖大碗茶的茶摊摆设也很简单，一张桌子，几条长凳即可。

Big Bowl Tea in Old Beijing

Big bowl tea was popular in old Beijing during 1930s and 1940s, which was developed from tea stands among people in ancient times. Big bowl tea used big pots or buckets to make jasmine tea. Then sellers would ladle it into coarse porcelain tea bowls with blue edge and cover the bowls with glass for sale. If passers-by felt thirsty, they would throw several cents, take the bowl and drink it. This big bowl tea could largely satisfy one's thirst, relieve the summer heat and people would feel refreshed. The point of this tea was not good quality water or tea, but just to satisfy one's thirst, suitable for common people. The tea stand itself was also simply decorated, with just one table and several benches.

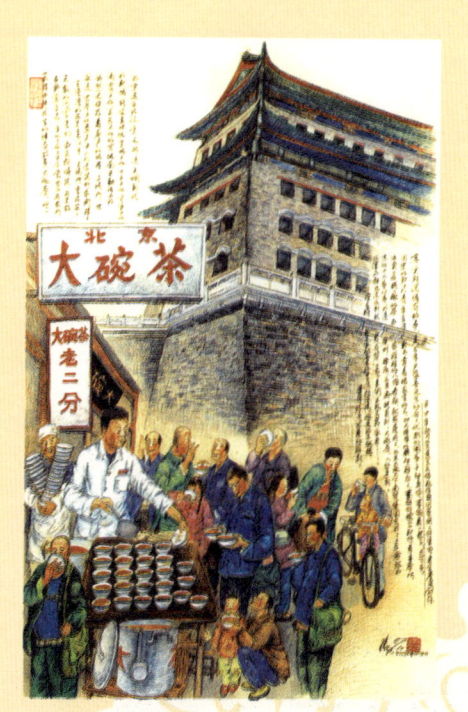

• 老北京的大碗茶
Big Bowl Tea in Old Beijing

> 紫砂茶具

紫砂茶具是用质地细腻、含铁量较高的特种黏土制成的，外表呈赤褐、淡黄或紫黑色的无釉精细陶质茶具。紫砂陶土含铁量高，有"泥中泥，岩中岩"之称。紫砂陶土质性特殊，表现为质地细

- 紫砂茶具
  Purple Clay Tea Set

> Purple Clay Teasets

Purple clay teasets, made of special fine clay which has high content of iron, are sorrel, light yellow or purple black delicate ceramic teaware without glaze. Purple clay contains a lot of iron and is called "the best of the clay, the best of the rock". Purple clay is very special, fine and bright. Since purple pottery needs mading under the temperature of 1000 ℃ -1200 ℃ , the production is very compact and will not leak. In other words, a change of the small particles on the smooth surface manifests "sand-like" effect with bubbles invisible to naked eyes. Purple clay teawares can absorb tea and keep its fragrance. In addition, they conduct heat slowly, so they will not scald hands. Even when there is a sudden change from cold to heat, they will not crack. Purple clay tea pot will make fragrant and warm tea, which does

腻，颜色鲜艳。由于成陶火温需在1000℃~1200℃之间，所以成品致密，不渗漏，即表面光滑平整之中含有小颗粒状的变化，表现出一种砂质效果，又有肉眼看不见的气孔。以紫砂陶所制成的茶具能吸附茶汁，蕴蓄茶味，且传热缓慢不致烫手，即使冷热骤变，也不致破裂。用紫砂壶泡茶，香味醇和保温性好，无熟汤味，能保茶之真髓。紫砂茶具用来泡乌龙茶、铁观音等半发酵茶最能展现茶味特色。

关于紫砂产生的年代，有不同的说法。有人认为早在宋代就已有紫砂器，但学术界比较认同紫砂起

- 供春制树瘿壶（明）
此壶高10.2厘米，宽19.5厘米，现藏于中国国家博物馆。
Purple Clay Pot in Burl Shape, Made by Gong Chun (Ming Dynasty, 1368-1644)
10.2 cm in height, 19.5 cm in width, it is now kept in the National Museum of China

not contain the smell of boiling water and keeps the essence of tea. They can best express the features of semi-fermented tea like oolong or *Tieguanyin*.

There are different versions concerning the first time people making purple clay tea pots. Some believe that they were made as early as in the Song Dynasty, yet scholars prefer the Ming Dynasty (1506-1521). In the Ming Dynasty, making loose tea promoted the development of purple clay pottery. At least, purple clay teawares from the mid-Ming Dynasty were found in the area of Yixing.

The first known master of purple clay pottery was Gong Chun during the Zhengde Period (1506-1521) of the Ming Dynasty. Gong Chun was a boy servant of scholars in his early years. It is said that when he was accompanying his master in *Jinsha* Temple, Yixing, he secretly learned how to make purple clay pot from an old monk. Later he used the clay precipitated on the bottom of the jar where the old monk washed hand. Then he made a pot which imitated the shape of a burl of a maidenhair tree next to the temple and engraved the pattern of the burl on the pot. The finished piece looked primitive, simple and lovely.

于明代说，明代散茶的冲泡直接推动了紫砂壶业的发展。至少在明代中期，宜兴一带的紫砂茶具已开始出现。

紫砂壶历史上第一个留下名字的大师是明正德年间（1506—1521）的供春。供春原姓龚，早年是一位官员的书童。传说他陪同主人在宜兴金沙寺读书时，寺中的一位老和尚很会做紫砂壶，供春就偷

Thus the pot became famous in a short time and people called it "the Pot of Gong Chun". Therefore, we can conclude that the development of purple clay pots from coarse handicraft to arts and crafts creation should be attributed to Gong Chun.

During the Wanli Period (1573-1620) of the Ming Dynasty, there were many master craftsmen of purple clay pottery, who had their own unique skills, especially Shi Dabin who represented the maturity of purple clay pottery making craft. At this time, there were three purple clay pottery modes and each mode had its masterpieces. The pots followed the artistic and humanistic features of the models of copper, tin utensils and Ming-style furnitures. This period was the first peak in the history of purple clay pot making. After the initial stage of prosperity in the Ming Dynasty, purple clay teaware saw its new peak. The Qing

- 紫砂海棠式提梁大壶（明）

现在能看到的最早的古代紫砂壶实物，是1965年出土于南京市中华门外马家山的明代太监吴经墓中的紫砂提梁壶。此壶高17.7厘米，口径7厘米，据该墓的砖刻墓志纪年推算，此物为嘉靖十二年（1533年）烧造。这是中国目前唯一有确切纪年可考的古紫砂器物。壶的年代久远，但制作工艺简单。

Big Purple Clay Pot with Handle in Begonia Shape (Ming Dynasty, 1368-1644)

The earliest purple clay pot that we find is a pot with handle, excavated from the tomb of Wu Jing, an eunuch in the Ming Dynasty, at Majia Mountain, Zhonghuamen, Nanjing in 1965. This pot is 17.7cm in height with a mouth of 7cm diameter. According to the engraved annals of this tomb, this pot was made in 1533. It is the only ancient purple clay utensil that has exact annals in China at the moment. It is old but its production process is simple.

偷地学。后来他用老和尚洗手后沉淀在缸底的陶泥，仿照寺旁大银杏树树瘤的形状做了一把壶，并刻上树瘿上的花纹，烧成之后非常古朴可爱，这壶一下子出了名，人们都叫它"供春壶"。可以说，宜兴紫砂壶从粗糙的手工艺品发展到工艺美术创作，应该归功于供春。

明代万历年间（1573—1620），紫砂名工辈出，各怀绝技，特别是时大彬的出现，标志着紫砂壶艺的成熟。紫砂壶的三大壶式这时已全面奠定，并均有上佳作品问世。这一时期紫砂壶的造型还较多地吸取铜锡器造型和明式家具的特点，从而积淀了更多的文化内涵和文人气息。可以说，万历年间是紫砂壶艺史上第一个鼎盛时期。经过明代的初步繁荣，到清代，紫砂茶具迎来了新的创作高峰。清代紫砂制作工艺大大提高，胎体细腻，制作规整，并出现了许多制造紫砂壶的名家。

此时紫砂壶品种日益增多，形制多姿多彩，泥料配色也更丰富，以朱泥、紫色为主体，还有白泥、乌泥、黄泥、梨皮泥、松花泥等多种色泽。制壶技艺、装饰手法也都

Dynasty witnessed a great increase in the skill of purple clay pottery making, which featured delicate pot base, orderly products and many master craftsmen.

There were more types of purple clay pots then, with different shapes and colors. Red and purple clay were mainly used with other colors like white, black, yellow, pear-skin and preserved egg. New skills in pot making and decoration emerged, so the royal palace also showed interest in purple clay utensils. They were made as tributes, and special pots with

- **杨彭年款描金山水诗句壶（清）**

  杨彭年是清代著名的制陶工艺家，制作紫砂壶的名家之一，活动于嘉庆、道光年间。

  Purple Clay Pot with Mountain, River and Poem Pattern in Gold, Made by Yang Pengnian (Qing Dynasty, 1616-1911)

  Yang Pengnian was a famous ceramic craftsman in the Qing Dynasty, one of the masters of purple clay pot making. He mainly made pots during the Jiaqing and Daoguang periods (1796-1850).

有新的创造和发明。由于技艺日益精进，紫砂器被宫廷皇室所看重，成为贡品，也因此出现装饰更为精美的特种工艺紫砂壶。

在紫砂壶上雕刻花鸟、山水和各种书法、绘画，器形也变成以几何型为主，尽可能扩大光洁面的面积，以便使用刻画装饰手段，表现文人所喜欢的书法、绘画、篆刻等内容。紫砂壶更多融入了文人的审美情趣，成为融文学、书法、绘画、篆刻于一体，既具有实用价值，又可欣赏、把玩的佳器。

more delicate decoration appeared.

Craftsmen engraved flowers, birds, mountains, rivers, calligraphy and paintings on the potteries. They chose geometric shapes, so that they could have larger smooth part to display what scholars prefer, such as calligraphy, paintings and seal cuttings. Purple clay pots carried more humane appreciation of beauty, which integrated literature, calligraphy, painting and seal cutting. They became both practical utensils and craftwork for appreciation.

紫砂泥的成分与类型

紫砂泥的主要矿物成分为石英、黏土、云母和赤铁矿。这些矿物的颗粒组成适中，矿物组成属于黏土—石英—云母系，特点是含铁量比较高。经过加工的紫砂泥熟土中含有石英砂和较高的铁量，呈紫色或紫红色，有极好的可塑性，烧成后质地呈细砂粒状，器表光滑，断面的细砂粒状看上去十分细腻。紫砂泥种类较多，制作紫砂壶的主要原料为紫泥、朱泥、绿泥。这三种紫砂泥产于不同的陶土矿层，经过加工，烧成后为"本色壶"。经工匠们的巧妙搭配后，其色彩非常丰富，成品可烧成海棠红、朱紫砂、葵黄、墨绿、白砂、淡墨、梨皮、豆青、新铜绿等几十种颜色，并且全凭原料呈现天然色泽，显得质朴高雅。

Component and Types of Purple Clay

The main mineral compositions of purple clay are quartz, clay, mica and hematite. They are evenly mixed, which can be described as "clay-quartz-mica" with high iron content. The processed purple clay contains more quartz and iron, which looks purple or purple red. This clay is very plastic. After being fired, the utensil will have fine sand and smooth surface and the cross section of the pottery looks very delicate. Purple clay comes in many types, among which purple, red and green clays are mainly used to make tea pots. These three clays are found in different seams of ceramic clay. After processing, the finished product is called "natural color pot". Under the ingenious arrangement of craftsmen, the potteries can have various colors, like begonia red, red purple clay, sun flower yellow, black green, white sand, light black, pear skin, bean green and new copper green. All these colors, plain and elegant, come from natural materials.

- **加工好的紫砂泥**

  一般情况下，50千克好的紫砂泥矿砂仅能提炼出3~3.5千克的好泥料，炼泥率不足10%。

  Processed Purple Clay

  Usually, 50kg quality purple clay can be extracted only 3-3.5kg good pug for use. The pug refinement rate is under 10%.

壶盖：紫砂壶烧成后，口和盖的配合应达到"直、紧、通、转"四项要求。"直"，是指盖的子口，要做得很直，举壶斟茶时，壶盖也不会脱出。"紧"，是指盖与口之间要做到"缝无纸发之隙"，严丝合缝。"通"，是指圆形的口和盖，必须圆得极其规正，盖合时要旋转爽利。"转"，是指方形（包括六方、八方）和筋纹形的口盖，盖合时可随意盖合，即可扣合严密，纹形丝毫无差。

The Lid: After finishing the pot, craftsmen should meet four requirements to make the lid and the mouth: straight, close, orderly and suitable. "Straight" means that the lid should be made in a very straight shape so that when serving tea, the lid will not fall off. "Close" refers to the tightness between lid and mouth, where even "a piece of paper or hair can not fit in". "Orderly" means the shape of the lid and mouth must be orderly and easy to turn. "Suitable" implies the tendon patterned square (or hexagon/octagon) lid should be able to open and close smoothly. In other words, they fit each other closely without change of the pattern.

壶嘴：壶嘴的制作非常讲究，嘴式的长短、粗细及安装位置都要恰当，壶嘴内壁必须光滑畅通，出水流畅，收水时不滴水、不流涎。壶嘴根部的出水眼因易被茶叶堵塞，从清代中期起做成网眼式。

The Spout: It requires great skills to make a good spout. The length, thickness and position of the spout should be appropriate. Its inner wall must be smooth and clear to let water out fluently. When the user stops pouring water, a good spout will not drop or drip. The root of the spout, where water comes out, is easy to be stuck by the tea leaves. Since the mid-Qing Dynasty, craftsmen have made it into a net shape.

壶把：壶把是为了便于执壶而设制的，有端把、横把、提梁三种基本形式。

The Handle: Made for people to carry the pot. The three main shapes are the level handle, the handlebar and the hoop handle.

壶体：壶体指紫砂壶的形体造型，是紫砂壶的主体。

The Body: The shaping of the purple clay pot, the main part of the pot.

器足：器足关系到紫砂壶的放置平稳性，其设计是否得当，直接影响紫砂壶的美观，故手工艺人对器足设计制作十分重视。

The Feet: The feet are closely related to the stability of tea pots and its design affects the look of the pot. Therefore, craftsmen pay great attention to the design of this part.

- 紫砂壶各部构造
  Constitution of a Purple Clay Pot

制壶名家时大彬

时大彬（1573—1648），明万历至清顺治年间的紫砂壶制作名家。在紫砂壶盛极一时的明朝，时大彬被称为"壶家妙手"，可见其制壶水平之高。时大彬的父亲时鹏是当时宜兴的紫砂壶名家之一，他从小深受父亲影响，经过勤奋钻研，他的制壶技艺远远超过了父亲，紫砂壶的艺术体系也从他这里走向了完备与成熟。

时大彬首创了在制壶泥料中掺砂的"调砂制壶法"，改进了紫砂壶泥片镶接成形的方法，使紫砂壶的制壶技艺获得了一个飞跃式的发展。紫砂壶原来器形很大，时大彬受到文人雅士的启发，将壶的器形由大改小，将文人情趣与制壶技艺融为一体。他制壶的态度十分严谨，如果作品不满意，马上毁掉，绝不吝惜。时大彬一生所制作品数以千计，但流传存世者极少。清乾隆年间，他的作品已被视为稀世珍宝。

Shi Dabin, the Pot Master

Shi Dabin (1573-1648) was a master of making purple clay pot during the Wanli Period of the Ming Dynasty to the Shunzhi Period of the Qing Dynasty. In the Ming Dynasty, when purple clay pots were extremely popular, Shi was named "magic pot maker", which showed his skill of making pots. Shi Peng, his father, was a master of making pot in Yixing at that time. Deeply influenced by his father and after hard study and work, Shi Dabin finally exceeded his father in pot making skills, and the art of purple clay pot making was raised to a more complete and mature level.

Shi Dabin was the first one to use "fit sand in making pot" method. By mixing sand into pug, it changed the way to join purple clay slices, which was a big step in pot making skill. Purple clay pots used to be very big. Shi Dabin was inspired by scholars and changed the size from big to small, which integrated literary feeling with pot making skill. He was serious at making pots. If the product was not satisfying, he would destroy it without hesitation. He made thousands of pots during his life time, but very few of them were passed down to later generations. In the Qianlong Period of the Qing Dynasty, his works had already been taken as treasures.

- 时大彬制紫砂开光方壶（明 万历）
Frame Outlining Square Purple Clay Pot, Made by Shi Dabin (Wanli Period of the Ming Dynasty)

紫砂壶的壶式

紫砂壶的式样主要有三大类：光货、花货、筋囊货，每一类又包括许多五花八门的壶式。

光货

光货是指壶身为几何体、表面光素的紫砂壶。光货有圆器、方器两大类。

圆器，即器的横剖面是圆形或椭圆形，圆器的轮廓由各种方向不同和曲率不同的曲线组成，讲究骨肉

- **圆器光货——孟臣款梨皮朱泥壶（晚清）**
Round Smooth Pot—Pear Skin Red Pot in the Meng Chen Style (Late Qing Dynasty)

停匀，比例恰当，转折圆润，隽永耐看，显出一种活泼柔顺的美感。

方器，即器的横剖面是四方、六方、八方等，方器的轮廓是由平面和平面相交所构成的棱线所组

Types of Purple Clay Pots

There are mainly three types of purple clay pots: smooth pots, engraved pots and tendon sack pots, each with many subtypes.

Smooth Pots

Smooth pots refer to geometric shape pots with smooth and plain surfaces. They can be divided into two types, round and square.

Round pots, the cross section of which is round or oval, have outlines combined by curves of different directions and curvatures. This kind of purple clay pots stresses an appropriate proportion between each part as well as the smoothness of curves. The pots stand detailed appreciation and manifest a lively and docile beauty.

Square pots, which may be rectangle,

- **方器光货——紫砂方壶**
Smooth Square Pot

成，讲究线面挺括平整，轮廓线条分明，展示出明快挺秀的阳刚之美。僧帽壶、传胪壶、觚棱壶等都是明清著名的方器壶式。

花货

花货又叫"塑器"，是以雕塑技法为制器的主要手段，讲究器形仿自然之形，惟妙惟肖，让使用者在沏茶时能体会到巧夺天工的美感。器形有三类：第一类是仿植物之形为器，如梅段壶、松段壶、竹段壶等；第二类是仿瓜果之形为器，如南瓜壶、佛手壶、藕形壶；

hexagon or octagon, have outlines combined by ridges of surfaces or intersections. This kind of pots stresses orderly lines, surfaces, and distinguished outlines in order to show lively and powerful masculine beauty. Mitral valve pots, summon pots and ladle ridge pots were famous pot types in the Ming and Qing dynasties.

Engraved Pots

Engraved pots, also called "carved pots", mainly use engraving skills. These pots focus on vivid imitation of natural items, so that people will feel their marvelous arts when making tea. They come in three types: firstly, those pots with plant-like shapes, like plum blossom pot, pine tree pot and bamboo pot; secondly, those with fruit shapes, such as pumpkin pot, chayote pot and lotus root pot; thirdly,

- 杨凤年制竹段壶（清）
Purple Clay Pot with Bamboo Handle, Made by Yang Fengnian (Qing Dynasty, 1616-1911)

- 邵大亨制鱼化龙壶（清）
Purple Clay Pot with "Fish Changing to Loong" Pattern, Made by Shao Daheng (Qing Dynasty, 1616-1911)

第三类是以动物之形为器，如鱼化龙壶，以动物之形为流、壶把的，也归于此类。此外，还有一些带浮雕装饰的紫砂壶，因装饰浮雕做得很显眼，也划归花货。

筋囊货

筋囊货又叫"筋纹器"，紫砂艺人把类似南瓜棱、菊花瓣等曲面形叫作"筋囊"，然后以这类"筋囊"为单元去构成壶形，并做到器表和器内一样，口部和壶盖的"筋囊"要上下对应、合缝严密，体现一种几何构图般的精巧和秩序之美。

those with animal shapes, like fish changing into loong pot, also those with an animal-shape handle. In addition, some purple clay pots have anaglyph. Since the anaglyphs on the pots are outstanding, these are also categorized in to engraved pots.

Tendon Sack Pots

Tendon sack pots are also called "tendon-line pots". "Tendon sack" refers to curves like pumpkin ridges or chrysanthemum pedals on the pots. Then craftsmen take these separate "tendon sack" to make a pot. They have to ensure the sameness between the out surface and the inner part, the close and fitted mouth and the lid. These tea pots show a geometric and orderly beauty.

- 菊花八瓣壶（明 万历）
  Octagonal Chrysanthemum Pedal Pot (Wanli Period of the Ming Dynasty)

- "鸣远"款菱花式壶（清）
  Water Caltrop Shape Pot of Chen Mingyuan Style (Qing Dynasty, 1616-1911)

紫砂壶的装饰技法
Technique of Ornament Purple-clay Tea Pot

• **紫砂加彩大茶叶罐（清）**

清代康熙年间，随着对外贸易的发展，为了迎合欧洲人崇尚华丽的审美习惯，紫砂艺人们开始在紫砂壶上涂彩，出现了彩釉装饰的紫砂壶，最初是施五彩，以后又出现珐琅彩、粉彩，统称为彩釉装饰。其中，乾隆年间的粉彩紫砂壶做工非常精细，可与瓷器媲美，是彩釉紫砂壶中质量最好的。

Colored Large Purple Clay Tea Jar (Qing Dynasty, 1616-1911)

In the Kangxi Period of the Qing Dynasty, with the development of export trade, purple clay craftsmen began to paint colorful patterns on the pots in order to meet the elegant European style. These were the first color glazed purple clay pots. At firstly, craftsmen only used polychrome; yet later, enamel glaze, famille rose were all in use, which were all called color glazed decoration. Among these works, famille rose glazed purple clay pots made in the Qianlong Period, were featured with delicate skills comparable to porcelains. They were the best of color glazed purple clay pots.

• **刻竹纹紫砂壶（清 杨彭年制 瞿应绍刻）**

紫砂艺人以握毛笔的姿势操刀，在泥坯上雕刻文字和图形的装饰手段，称为"陶刻装饰"，可分为刻底子和空刻两种。刻底子是指先用毛笔在坯体上书画，定稿后依墨迹镌刻；空刻则是由擅长书画的陶刻艺人拿刀直接镌刻。由于文人及书画家们积极参与陶刻装饰，陶刻装饰成为了紫砂装饰的主流。

Purple Clay Pot with Engraved Bamboo Pattern (Made by Yang Pengnian, engraved by Zhai Yingshao, Qing Dynasty, 1616-1911)

"Ceramic engraved decoration" refers to the skill that craftsmen hold the writing brush in a writing style and engrave characters and patterns on the clay base. This skill includes carving on the draft and carving directly. In draft carving, craftsmen first draft patterns on the base with brushes and then carve according to the draft. Yet in undrafted carving, those craftsmen who are good at painting and calligraphy carve directly with knives. Since scholars, calligraphers and painters all actively take part in engraved decorating, and ceramic decoration become the mainstream of purple clay decoration.

各种材质的老茶具
Ancient Teawares of Different Materials

- 紫砂刻花诗句壶（20世纪上半叶）

紫砂壶的刻画装饰是制壶人在署名刻款的基础上发展起来的，也与文人书画家参与紫砂壶设计有关。在著名的书画家陈鸿寿的大力倡导下，壶身一面镌刻壶铭，另一面刻画绘画，已成为定式，使紫砂壶具有浓郁的文人情趣和风格。

Purple Clay Pot with Engraved Poem Pattern (First Half of the 20th Century)

Engraved decoration on purple clay pots was developed from pot makers' signature engraved on the pot, and it was also scholars' contribution to the design of the pots. With the promotion of famous painter Chen Hongshou, it became a pattern of engraving poems on one side of the pot and paintings on the other. This made these pots carry more scholar style and feeling.

- 时大彬款紫砂雕漆四方壶（明）

雕漆装饰是以紫砂壶为内胎，在紫砂壶表面髹几十道大漆，再用刻刀在漆层上剔刻出繁复精致的花纹图案。雕玲珑是指在壶的外围做一圈透雕装饰，显得玲珑剔透，十分精致。

Square Purple Clay Pot with Engraved Lacquer of Shi Dabin Style (Ming Dynasty, 1368-1644)

Engraved lacquer decoration refers to the skill that craftsmen apply dozens of layers of lacquer on the surface of the pot and carved complicated and detailed patterns on the lacquer layer. Engraving *Linglong* is the skill that craftsmen make a round openwork carving on the pot, which looks very exquisite and delicate.

- 紫砂炉均加彩大方壶

"炉均"也是一种釉彩装饰，是在紫砂胎周身施满低温铅釉，这样烧成后，在匀净的天蓝色釉面中有细腻而致密的白毫，很有特色。

Evenly Fired Color Glazed Large Square Purple Clay Pot

"Evenly-fired color glaze" is a kind of glazed decoration. Craftsmen apply low-temperature glaze all over the pot base. When fired, the pot will show very special sky blue evenly glazed with delicate and dense pekoe within.

- 紫砂如意云纹荷花茶壶

将纹样的一个单元制成模具,用模印法制好装饰纹样泥片,再粘贴在紫砂壶坯上,形成一种结构严密、纹样统一的装饰效果,这种方法叫做"印贴装饰"。常见纹样有如意云纹、蕉叶纹、蝉纹、夔纹、龙凤纹、水波纹等。

Purple Clay Pot with *Ruyi*-shaped Cloud and Lotus Pattern

The skill of "stamp and stick decoration" means making molds of parts of the pattern and stamp clay slices with patterns, then sticking the clay slices to the pottery base. Thus, the decoration will be structured closely together and have a unified pattern. The most common patterns are the *Ruyi-shaped* cloud, the banana leaf, the cicada, *Kui* (one-leg monster in fable) and water waves.

- 紫砂描金方壶(清)

在紫砂胎上描金是先在描金的纹样处涂上一层底釉,用750℃－800℃的温度烧成,再蘸金水在釉纹上描画,然后再烧烤一次。由于工序繁杂,成本很高,因此历史上此类作品不多。

Square Purple Clay Pot with Gold Caligraphy (Qing Dynasty, 1616-1911)

Gold tracing on purple clay means that craftsmen apply glaze on the design pattern and fire the base to 750℃-800℃. Then they apply liquid gold to the glazed pattern and fire it again. Due to the complicated process and high cost, there are not many works of this kind.

- 紫砂泥绘壶(明)

用本色泥料或白泥、朱砂泥、乌泥等制成泥浆,用毛笔蘸泥浆在尚有一定湿度的壶坯体上绘画花鸟或山水,这一技法被称为"泥绘",因画面有一定的厚度,犹如薄浮雕,又叫"堆画"。

Purple Clay Pot with Clay Painting (Ming Dynasty, 1368-1644)

The skill of using original color clay or a mixture of white, red and black clays to paint flowers, birds, mountains and rivers on a wet pottery base with the writing brush is called "clay painting". Since the paintings stand out like anaglyph, they are also named "piled painting".

113

各种材质的老茶具 Ancient Teawares of Different Materials

工夫茶具四宝

　　工夫茶流行于中国的福建、广东、台湾等地区，尤其是在闽南、潮汕地区，几乎家家都有工夫茶具，男女老少都会泡工夫茶，即使寄居他乡或移民海外，也保留着这种品茶习惯。工夫茶指的不是茶叶，而是一种冲泡茶的程式和方法，起源于明代，在清代发展成熟，进入鼎盛期。工夫茶以茶具精致小巧、冲泡考究为主要特点，一般不用红茶和绿茶，而用半发酵的乌龙、铁观音等，且其茶叶远没有茶具讲究。工夫茶的茶具往往是"一式多件"，一套茶具一般以十二件为常见，其中有四件必不可少的茶具，称为"四宝"。它们分别是：孟臣冲罐、若琛瓯、玉书锅、红泥烘炉。

　　孟臣冲罐是用宜兴紫砂陶制成的小茶壶。传说孟臣是明代江苏宜兴紫砂壶名匠，姓惠，擅长制作小壶。这种小壶用于泡茶时，色香味皆蕴，壶经久耐用，即使以沸水注入空壶也有茶味，盛夏隔夜茶不易馊。茶壶耐热性强，寒冬沸水注入，无爆裂之虞。另外，此壶传热性慢，用时不烫手，使用越久越有光泽，显得古色古香。

　　与"孟臣"合称茶具双璧的是若琛瓯。这是一种薄瓷小杯，薄如纸，白似雪，小巧玲珑，酷似半个乒乓球，一只杯仅能容七八毫升茶汤。容积如此之小，主要是因为工夫茶多为闲暇时品饮，而非为解渴。古代正宗的若琛瓯产于江西景德镇，杯底有"若琛珍藏"字样。

　　泡饮工夫茶对水及水壶的选择相当严格，"玉书锅"是一种工夫茶专用的薄瓷水壶，不但避免了金属水壶的异味，且保温性好，冬日里离炉许久仍能保持水温，久用也不易结水垢。相传，古时有位工匠设计出此壶后，邀来三五茶友为它命名。茶友见此壶烧出的水清洁如玉，倒水宛若玉液输出，就取名"玉输"。后人认为"输"字不祥，便用"书"字代替。

　　红泥烘炉，是选取粤东优质的高岭土烧制，高尺余（30~40厘米），置炭的炉心深且小，能使火势均匀且省炭。小炉有门有

▶ 紫砂小壶
Small Purple Clay Pot

盖，茶人喜用橄榄核为燃料，火焰温度高而无杂味。这种炉子通风性能好，即使有水溢入，炉火也不致熄灭。有的炉门处还刻有跟茶有关的对联，显得古朴雅致。

Four Treasures of *Gongfu* Teaware

Gongfu tea is very popular in Fujian, Guangdong and Taiwan, especially in the southern part of Fujian Province and Chaozhou, Shantou area, where almost every family has a *Gongfu* tea set. Men and women, old and young, all know how to make *Gongfu* tea. Even after they move to other places, these people will still keep this habit. *Gongfu* tea does not refer to tea leaves, but a way of makeing tea. It originated in the Ming Dynasty, matured and reached its peak in the Qing Dynasty. *Gongfu* tea is featured with delicate tearwares and special making method. People seldom use black or green tea to make *Gongfu* tea. Instead, oolong tea and *Tieguanyin* are more preferred and they are far less delicate than the teaware itself. *Gongfu* teaware is always in a "one set and several wares" style. One tea set often has twelve pieces, among which four parts are necessary, called "four treasures". They are *Mengchen* pot, *Ruochen* cup, jade fluid cauldron and red mud stove.

A *Mengchen* pot is the small tea pot made of purple clay from Yixing. It is said that Mengchen was a purple clay pot master from Yixing, Jiangsu Province in the Ming Dynasty. His surname was Hui and he was good at making small pots. These small tea pots could maintain the smell, fragrance and look of the tea. In addition, his pots stand wear and tear. Even when just boiling water are added in the pot, people can still smell the fragrance of tea, and tea seldom becomes sour when kept in the pot overnight. The pot is heat-resisting and will not explode even when receiving boiling water in freezing winter. Besides, it conducts heat slowly, which can prevent the holder's hand from being scalded. The longer it is used, the brighter and more classic beautiful it will look.

A *Ruochen* cup is the best match with the *Mengchen* pot. It is a small, delicate cup, as thin as paper, as white as snow, looking like a half table tennis ball, which can just take seven or eight milliliters tea. The design of such a small volume is mainly for the purpose of relaxing by having *Gongfu*

• 工夫茶茶杯
Gongfu Tea Cups

tea, not for quenching thirst. Authentic ancient *Ruochen* cups were made in *Jingdezhen* Kiln, Jiangxi Province, with "collection of Ruochen" on the bottom of the cups.

Gongfu tea has high requirement for water and pot. "jade fluid cauldron" is the thin porcelain cauldron made for *Gongfu* tea. It not only avoids the odd taste of metal cauldrons, but also is good at keeping warmth. In winter time, it can keep the temperature of the water for a long time after leaving stove. Even when used for some time, it will not take scale. It is said that an ancient craftsman invented this cauldron and invited several friends to give it a name. People found that the water coming out from the cauldron was as clean as jade, so they named it *Yushuguo* (jade fluid flowing out from the cauldron). However, later people thought that the character *Shu* (flowing out) also means "to lose" in Chinese, which was not auspicious. Thus, they changed this character into a homonymous character, which means "book", so it literally means "jade book cauldron".

The red mud stove made of high quality kaolin clay of eastern Guangdong, is 30-40 cm tall. The core of the stove is deep but small; so it can keep an even fire and save charcoal at the same time. This stove has a ventilation door and a cover. Tea makers love to use olive pips as fuel, which produces heat without unpleasant smell. This stove has good ventilation. Even when water comes in, the fire will not extinguish. Some stoves have engraved couplets about tea on the ventilation door, looking siple yet elegant.

- 玉书锅和红泥烘炉
  Jade Fluid Cauldron and Red Clay Stove

> 金银茶具

　　金银在中国古代一直被视作贵重之物，中国人早在三千多年前的商代就已使用金银器。金银茶具按质地分类，以银为质地者称"银茶具"，以金为质地者称"金茶具"，银质而外饰金箔或鎏金称"饰金茶具"。金银茶具大多以锤成型或浇铸焊接，再施以刻画或镂空装饰。由于金银延展性强，耐腐蚀，又有美丽的色彩和光泽，加上成品极为精致，因此非常昂贵，多专供帝王、富贵之家使用。

　　唐代是我国金银器发展史上的第一个高峰期。唐代的开放程度很高，对外交流也频繁，工匠们吸收了西亚一带的金银器加工工艺，并结合本土的工艺，把金银器制作推向一个创作的高潮。此外，唐代时

> Gold and Silver Teawares

Gold and silver are always taken as precious items in ancient China. In the Shang Dynasty about 3,000 years ago, Chinese people had already used gold and silver utensils. In terms of materials, there are teawares made of silver and gold, and also silver tea sets covered with gold foils called "gold decorated tea sets". Most gold and silver teawares are hammered in to shape, or die-casted, or welded, and decorated with engraved or hollowed out patterns. Due to the good ductility and corrosion resistance of gold and silver, their beautiful colors and gloss, gold and silver tea sets are extremely delicate, very expensive, and mostly used by royal or rich families.

　　The Tang Dynasty witnessed a peak in the history of gold and silver utensils. It was a relatively open society then and had more frequent communication with

大量的金银矿被开发出来，故金银器的加工工艺也有了很大的突破。从现存及出土的资料来看，唐代的金银茶具不少，其中最具代表性且等级最高的，是陕西省扶风县法门寺地宫出土的鎏金银茶具。这些茶具制作十分考究，质地精良，它们是

other countries; craftsmen learned gold and silver processing skills from Western Asian areas, combined these skills with traditional local ones and improved the quality of gold and silver utensils. In addition, many gold and silver ores were discovered in the Tang Dynasty, which was also led to a big breakthrough in

- 鎏金银笼子（唐）

通高17.8厘米，口径16厘米，腹深10.2厘米，制作精美，是用来装放茶饼的。由于唐代茶叶以饼茶为主，茶饼易受潮，要用纸或叶包装好放在茶笼里，挂在高处，通风防潮，饮用时取出。皇家用金银制作的茶笼子讲究精工细作，尽显皇族贵气。

Silver Cage Gilded with Gold (Tang Dynasty, 618-907)

This exquisite cage is 17.8cm in height, 16cm in diameter and 10.2cm in depth, made for storing tea cakes. Tea leaves were mainly made into tea cakes and they would easily absorb moisture, so they needed covering with paper or leaves in the cage. The cage should be hung in a higher position in order to be well ventilated and kept dry. Then people could take them out when needed. These tea cages for royal families had high requirements for making skills so as to show their noble deposition.

- 鎏金摩羯纹银盐台（唐）

鎏金摩羯纹银盐台，由盖、台盘、三足架三部分组成。通高25厘米，盖做成卷荷形状，十分精致。从台盘支架上的錾文可知，其用途是盛放盐巴，该盐台成为唐代有饮茶时添放盐花的习惯的有力佐证。

Silver Salt Stand Gilded with Gold Makara Pattern (Tang Dynasty, 618-907)

This salt stand was made up by cover, plate and a tripod. The 25cm high stand has a delicate cover shaped like a curly lotus leaf. From the engraved characters on the tripod, we can see that this stand was made for storing salt—evidence that people in the Tang Dynasty liked to put salt in tea.

唐咸通十五年（874年）封存的，到出土时已有1120年的历史，由此可证明中国唐代饮茶之风的盛行。

宋代的金银器与唐代相比则显得更加普及，除皇室宫苑有专门加工金银器的机构外，城里已有出售金银器皿的专门店铺，一些富商巨贾及家底殷实的家庭都可以使用金银制品。据《东京梦华录》记载，一些民间茶肆酒楼的饮具也用金银

the processing skills of gold and silver utensils. According to the materials that we have at the moment, there were many gold and silver tewares in the Tang Dynasty. The most typical and high-ranking ones among them are the silver tea sets gilded with gold excavated from *Famen* Temple, Fufeng, Shanxi Province.

These tea sets were perfectly made with high-quality materials. They were sealed in the 15th year of Xiantong Period (874) of the Tang Dynasty,

- 鎏金银茶罗子（唐）

Silver Tea Sieve Gilded with Gold (Tang Dynasty, 618-907)

鎏金银茶罗子，分罗框和罗屉，同置于方盒内，罗框长11厘米，宽7.4厘米，高3.1厘米。罗屉长12.7厘米，宽7.5厘米，高2厘米。茶饼在茶槽中碾碎成末，尚需过罗筛选。茶罗子是晚唐兴起的点茶的关键茶具。

This tea seive has a box and a drawer both in the square case. The box is 11cm in length, 7.4cm in width and 3.1cm in height. The drawer is 12.7cm in length, 7.5cm in width and 2cm in height. After tea cakes are ground into powder in the slot, the powder needs filtering with the sieve. Tea sieve was a key tea ware in late Tang Dynasty.

- 鎏金茶槽子和茶碾子（唐）

鎏金茶槽子，通高7厘米，最宽处5.6厘米，长22.7厘米。鎏金茶碾子轴长21.6厘米，碾轮边窄，轮径8.9厘米。由于唐代流行品饮末茶，凡饼茶需要用茶碾碾成粉末，该茶槽子和茶碾子就是用来碾茶的。

Tea Slot and Tea Roller Gilded with Gold (Tang Dynasty, 618-907)

Tea slot gilded with gold is 7cm in height, 5.6cm in the width and 22,7cm in length. The axle of the roller is 21.6cm with a sharp roller wheel, which is 8.9cm in diameter. As people of the Tang Dynasty preferred to drink tea powder, all the tea cakes needed grinding, which was the function of this tea slot and roller.

制作。而宋人所使用的茶具也以金银为上品，视其为身份和财富的象征。明代金银制品技术并无多少创新，但明代帝王陵墓中出土的金银茶具文物却精美无比，如埋葬万历皇帝的定陵出土的玉碗，其碗盖及碗托均为纯金錾刻而成。清代金银器工艺空前发展，皇家金银茶具更为普遍，史料曾记载太监用玉碗、金托、金盖的茶具伺候慈禧太后。

showing the popularity of tea drinking of the time.

Gold and silver utensils became more popular in the Song Dynasty. Besides the imperial court which had special gold and silver processing unit, shops selling them could also be found within the city. Rich businessmen and families could also use them. According to *Reminiscences of the Eastern Capital*, some restaurants among ordinary people also used gold and silver drinking wares. People of the Song Dynasty took gold and silver tearwares as high standard ones and the symbol of social status and wealth. In the Ming Dynasty, the skills of making these utensils were not largely improved, though the gold and silver tearwares excavated from the emperor tombs of the Ming Dynasty were very delicate. For example, the jade bowl from the Ding Mausoleum of Emperor Wanli had pure gold engraved patterns on cover and base. A great development was seen in the Qing Dynasty, and it became more common to find them in the royal palace. According to historic records, eunuches used jade bowls, gold bowl bases and gold covers to serve Empress Dowager Cixi.

- 鎏金银茶匙（唐）
  鎏金银茶匙，长19.2厘米，柄长而直，匙面平整。
  Silver Tea Spoon Gilded with Gold (Tang Dynasty, 618-907)
  This spoon is 19.2cm long, with a long straight handle and a flat scoop.

- 鎏金刻梅花纹银碗（宋）
  Silver Bowl with Gilded Plum Blossom Pattern (Song Dynasty, 960-1127)

- 鎏金银汤瓶（宋）
Silver Tea Pot Gilded with Gold (Song Dynasty, 960-1127)

- 镀金镂花银茶壶
Gilt Ornamental Engraving Silver Tea Pot

- 鎏金银荷叶托盏（宋）
Silver Bowl and Base Gilded with Gold Lotus Pattern (Song Dynasty, 960-1127)

- 金茶碗（明）
Gold Tea Bowl (Ming Dynasty, 1368-1644)

各种材质的老茶具
Ancient Teawares of Different Materials

> 琉璃茶具

琉璃，就是玻璃，古人又称为"琉琳"，一般是用含石英的砂子、石灰石、纯碱等混合后，在高温下熔化、成形，再经冷却后制成。在大约3100多年前的西周时代，中国人就初步掌握了琉璃制造技术。经测定，中国古代的琉璃成分以铅钡为主，与西方的钠钙玻璃分属两个不同的系统。钠钙玻璃耐

- 琉璃盏托（唐）
Glass Bowl and Base (Tang Dynasty, 618-907)

> Glass Teawares

Liuli, or glass, was also called *Liulin* by ancient people. It is usually made with a mixture of sand, quartz, limestone and soda ash. These materials are melted in high temperature, formed in shape and cooled down. In the Western Zhou Dynasty, about 3,100 years ago, Chinese people had already grasped the basic skills of making glass. It is tested that ancient Chinese glass mainly contained lead and barium, which is different from western soda lime glass. Soda lime glass is good at heat resistance and flexibility, while lead barium glass made in low temperature is not suitable to make eating utensils. With the increase in communication with other countries and the import of western glass products, Chinese craftsmen began to make colored glass teawares in the Tang Dynasty. The light yellow plain glass

• 琉璃茶壶
Glass Tea Pot

温性能较好，适应性较强；而铅钡玻璃烧成温度较低，不适合制作饮食器皿。直到唐代，随着中外文化交流的增多，西方琉璃器的不断传入，中国才开始烧制琉璃茶具。陕西扶风法门寺地宫出土的由唐僖宗供奉的素面圈足淡黄色琉璃茶盏和素面淡黄色琉璃茶托，就是地道的中国琉璃茶具，虽然其造型原始、装饰简朴、质地浑浊、透明度低，但却表明中国的琉璃茶具唐代已经起步，在当时堪称珍贵之物。宋代时，中国独特的高铅琉璃器具相继问世。元、明时，规模较大的琉璃作坊在山东、新疆等地出现。清康熙时，在北京还开设了宫廷琉璃厂，只是自宋至清，虽有琉璃器件生产，但多以生产琉璃艺术品为主，且身价名贵，只有少量茶具制品，始终没有形成琉璃茶具的规模生产。

tea bowl with round foot and its base, excavated from the underground palace of *Famen* Temple presented by Emperor Xizong of the Tang Dynasty in Shanxi Province, is a traditional Chinese glass teaware. Although they were featured with a primitive style, plain decoration, cloudy quality and low transparency, this showed that people of the Tang Dynasty had begun to make and treasure glass tewares. In the Song Dynasty, craftsmen made unique Chinese glass utensils with high lead content. In the Yuan and Ming dynasties, large scale glass workshops could be found in Shandong and Xinjiang. In the Kangxi Period of the Qing Dynasty, there was also a royal glass plant in Beijing. However, from the Song Dynasty to the Qing Dynasty, although there were many glass products, most of them were artworks and very expensive. There were only a small number of glass teawares produced.

> 漆茶具

漆器是采天然漆树汁液进行炼制，掺进所需色料而制成的器物。中国漆器起源甚早，在7000年前的河姆渡文化遗址中就发现了木胎漆碗。殷商时代，人们已经懂得在漆液中掺入各种颜料，或者在漆器上贴金箔、镶上松石。到了唐代，由于瓷器的发达，漆器已经往工艺品方向发展。宋代时，漆器被分成两大类，一类比较粗放简朴，光素无

> Lacquer Teawares

Lacquer utensils are made with refined natural lacquer tree sac and pigments. Chinese people started making lacquer utensils very early. A wooden base lacquer bowl was found out in the Hemudu culture relics, which was from 7000 years ago. In the the Shang Dynasty, craftsmen had already known how to add pigments in the lacquer, or inset gold foil or tophus on the lacquer utensils. In the Tang Dynasty, due to the development of porcelains, lacquer utensils began to evolve to artworks. In the Song Dynasty, lacquer utensils were divided into two categories. One kind was plain, simple and mainly used by ordinary people while the other kind was featured with delicate design and skill. Some even used

- 漆盏托（宋）
Lacquer Bowl and Base (Song Dynasty, 960-1279)

纹，多为民众所用。另一类则精雕细作，工艺奇巧，有的甚至用金银做胎，较为名贵。清代乾隆年间，福州人沈绍安用脱胎漆工艺制作茶具。漆器茶具乌润轻巧，光彩夺目，又融书画艺术于一体，与其说是茶具，还不如说是艺术品。

gold or silver as bases, which were very precious. In the Qianlong Period of the Qing Dynasty, Shen Shao'an from Fujian Province created teawares with bodiless lacquer. Lacquer teawares were light, handy, bright and colorful while also combined with calligraphy and paintings. Therefore, it was better to take them as artworks rather than teawares.

- **御题诗剔红碗（清 乾隆）**

  剔红是雕漆品种之一，就是在器物的胎型上，涂上几十层朱色大漆，待干后再雕刻出浮雕的纹样。此技法成熟于宋元时期，发展于明清两代。

  Red Lacquer Bowl with Carved Emperor's Words (Qianlong Period of the Qing Dynasty)

  Red carved lacquer, a kind of carved lacquers. Craftsmen applied dozens of layers of red lacquers on the base; when they got dry, anaglyph patterns were carved. This skill became mature in the Song and Yuan dynasties and developed in the Ming and Qing dynasties.

- **描漆牡丹纹盏托（清）**

  描漆就是在光素的漆地上用各种色漆画出花纹的技法，又称"彩漆""描彩"。清代雍正时期作品尤其精美。

  Painted Lacquer Bowl and Base with Peony Decorations (Qing Dynasty, 1616-1911)

  Lacquer painting, a skill of drawing colorful patterns on plain lacquer. It is also called "colorful lacquer" or "sketched colors". The products made during the Yongzheng Period of the Qing Dynasty were extremely outstanding.

脱胎漆器

　　脱胎漆器是一种极富民族特色的手工艺产品，与北京的景泰蓝、江西的景德镇瓷器并称为中国传统工艺"三宝"，享誉国内外。制作脱胎漆器工艺品的工序非常复杂，先以泥土、石膏做成坯胎，阴干后涂上生漆，用捶打过的麻布和生漆在坯胎上逐层裱褙，一般麻布和生漆要裱褙四层，用泥胎和石膏固定后在底部打个圆孔，待漆干后放在水中浸泡，直到泥土和石膏的内胎脱去，留下漆布器形。脱胎制作完后，再经过上灰、打磨、上漆修整、推光等十几道工序，并施以各种装饰图案，这样光亮绚丽的脱胎漆器便完成了。脱胎漆器的装饰技法具有绘画和雕刻工艺的双重属性。色泽主要有红、黑两种，再加以巧妙的调配增加黄、绿、蓝等颜色。

Bodiless Lacquer Utensils

Bodiless lacquer utensil is a craftwork with national features. It is one of the "three treasures" of traditional Chinese craftwork, along with Cloisonné from Beijing and porcelain from Jingdezhen, Jiangxi Province, which are well-known throughout the world. The process of making bodiless lacquer craftwork is very complicated. Firstly, craftsmen need to make pottery bases with clay and raw plaster and apply lacquer after the bases are dry. Then, they will mount about four layers of sackcloth hammered heavily and raw lacquer on the base. Later, they will fix them with unfired pottery and plaster, drill a hole on the bottom. When the lacquer is dry, they put the pottery in the water until the inner base with clay and plaster peel off and leave only the model. After this stage, craftsmen still need to do a dozen processes like adding lime, grounding, applying lacquer, trimming, brightening and painting various kinds of patterns on it. Thus a colorful and bright bodiless lacquer utensil is finished. The skill of making this utensil combines both painting and carving. It mainly uses red and black, added with colors like yellow, green or blue.

> 锡茶具

锡茶具的兴起是明清茶具的一个重要特点。锡金属的使用在中国的商周时期就开始了，但当时只是把锡作为合金材料加工。锡器发展的黄金时期应该是明清两代。明代锡茶壶与紫砂壶一样，受到文人雅

- 锡胎椰壳雕茶叶瓶（清）
  Tin Base Tea Bottle with Carved Coconut Shell (Qing Dynasty, 1616-1911)

> Tin Teawares

The emergence of tin teawares is characteristic of the Ming and Qing dynasties. Early in the Shang Dynasty people began to use tin, yet it was just a material for alloy process. The golden era of tin utensils are the Ming and Qing dynaties. Tin tea pots like purple clay tea pots in the Ming Dynasty were welcomed by scholars, and there appeared many masters in making them, and many splendid scholarly tin pots were made regardless of labor or money.

In the Jiajing Period of the Ming Dynasty, Zhao Liangbi from Suzhou learned from the purple clay pot and then made some great tin pots. After him, Master Gui Fuchu used raw tin to make pottery, ebony to make handle and jade to make spout and lid. These tea pots were very expensive even at that moment. The tin teawares made by Wang

士的推崇，一时间锡壶制作名家辈出，他们往往不惜工本，制作出许多精美绝伦的文人锡壶。

明代嘉靖年间，苏州人赵良璧借鉴紫砂式样，制作锡器。其后的制锡名家归复初，以生锡制壶身，用檀木做壶把，以玉做壶嘴和盖顶，他的作品在当时卖价就很高。嘉兴人王元吉制作的锡茶具以精巧著称，色泽似银，壶盖和壶身十分严密，合上之后，提盖而壶身亦

Yuanji from Jiaxing were famous for its delicate feature. The teawares looked like silver. The lid and body were connected very well. When you lifted the lid, the whole pot would go up. This was similar to the purple clay tea pots made by Shi Dabin. Tin teawares were still used the Qing Dynasty and this period witnessed several skillful masters in making tin utensils. The most famous one at the beginning of the Qing Dynasty was Shen Cunzhou. He was good at making tin

- 锡茶叶瓶（清）

用锡器储存茶叶，可以使茶叶长期储存而不变味、不变色。从古至今，许多喜欢喝茶的人都爱使用锡罐储茶。

Tin Tea Leaves Bottle (Qing Dynasty, 1616-1911)

Tin utensils can maintain the smell and color of tea leaves for a long time. From ancient times till now, tea lovers all like to use tin utensils to store tea.

- 朱坚款锡茶壶（清）

朱坚是晚清时期的制壶高手。这把锡茶壶呈浑圆瓜形，盖上镶碧玉纽，壶把镶白玉，壶腹一面刻盛开的梅花图案，另一面刻有诗文。

Tin Tea Pot of Zhu Jian Style (Qing Dynasty, 1616-1911)

Zhu Jian was a pot making master in the late Qing Dynasty. This tin pot was in a melon style, with jade button on the lid and jade on the handle. One side of the pot body was engraved with blooming plum blossoms while the other side was carved with poem.

起，与时大彬的紫砂壶特点相同。锡制茶具在清代继续使用并流行，涌现出不少具有高度艺术修养的锡器制作高手。清初最有名的要数沈存周，他善制各种式样的锡茶具，对壶形把握很准确，其制锡壶包浆水银色，光可鉴人，所雕刻的诗句、姓氏、图印均规整精良。道光、咸丰年间还出现了王善才、刘仁山、朱贞士等锡器制造名手，所制锡器也极为精工。

sets of various types and designing the shapes of the pots. The tin pots he made have water mark color, which was bright and shinning. All the poems, names and paintings on the pots were also very orderly. In the Daoguang and Xianfeng periods, masters like Wang Shancai, Liu Renshan and Zhu Zhenshi all made many delicate tin utensils.

- **杨彭年制紫砂镶玉锡包壶（清）**

  镶玉锡包壶，将壶外周以锡片包镶，再在锡片上刻画诗文绘画，嘴口、盖纽及壶把均用玉镶接，制成后显得高雅别致，可惜的是，这种特殊装饰技法已经失传了。

  Purple Clay Pot with Tin and Jade Cover, Made by Yang Pengnian (Qing Dynasty, 1616-1911)

  This pot was covered with engraved poems and paintings on tin. The pot spout, lid button and handle were connected with jade, looking unique and elegant. Unfortunately, this special decoration skill did not pass down.

> 珐琅茶具

珐琅器，就是以珐琅为装饰而制成的器物，按照胎骨的不同可以分成金属胎、瓷胎两类。一般所说的"珐琅器"，指的就是金属胎珐琅器，按照工艺的不同又可分为掐丝珐琅器、画珐琅器、錾胎珐琅器和透明珐琅器几个品种。其

> Enamel Teawares

Enamel utensils refer to utensils decorated with enamel. In terms of different bases, enamel utensils can be categorized into metal base and porcelain base. The "enamel utensils" are actually metal base enamel utensils. In terms of different making skills, they can also be grouped into filigree, painted, carved base and transparent enamel utensils. Among these types, filigree and painted enamel skills are the most common ones.

Copper filigree enamel utensil was first made in the Yuan Dynasty and reached its peak time in the Jingtai Period of the Ming Dynasty (1450-1456). Since the enamel utensils at that time were mainly in blue, they were also named

- 铜胎画珐琅提梁壶（清）
Copper Base Enamel Pot with Hoop Handle (Qing Dynasty, 1616-1911)

• 景泰蓝盏托
Cloisonné Bowl with Base

中，以掐丝珐琅和画珐琅工艺的茶具较为常见。

　　铜胎掐丝珐琅器的制造历史可追溯到元朝，在明景泰年间（1450—1456）最为盛行，又因当时多用蓝色，故又名"景泰蓝"。景泰蓝以紫铜作坯，制成各种造型，再用金线或铜丝掐成各种花，中间填充珐琅釉，经烧制、磨光、镀金等工序制成。景泰蓝茶具造型特异，制作精美，图案多样，色彩富丽，金碧辉煌，具有鲜明的民族特色。而画珐琅又称"洋瓷"，是用珐琅釉料直接在金属胎上作画，经烧制而成。画珐琅茶具既具有景泰蓝厚重、端庄的特色，又有瓷器明丽、清雅的风采，艺术特征独具一格。珐琅茶具大多为盖碗、盏托、茶壶等，制作精细，花纹繁缛，气派华贵。

"*Jiangtailan* (cloisonné)". Cloisonné used purple copper as its base to make various types and added with gold or copper filigrees. It was filled with enamel glaze and fired, polished and glided before finished. Cloisonné tewares had their unique features with delicate design, diversified patterns, bright colors, splendid look and typical national characteristic. Painted enamel utensils were also called "foreign porcelains", where craftsmen drew paintings directly on the metal bases with enamel glaze before firing. Painted enamel teawares had the dignity of cloisonné as well as the bright elegance of porcelains, which were very special in art style. Most of the enamel teawares were bowls with cover, bowls with base or tea pots. They were delicately made with a great variety of patterns and a feeling of luxury.

> 玉石茶具

　　玉石茶具是指用玉石雕制的饮茶用具。狭义的玉材有软玉（如羊脂玉）和硬玉（如翡翠）两大类，而广义的玉石包括硬玉和软玉、蛇纹石、绿松石、孔雀石、玛瑙、水晶、琥珀、红绿宝石等多种彩石玉。

　　中国玉器工艺历史悠久，唐时

> Jade Teawares

Jade teawares refer to those made by carved jade. Narrowly speaking, jade includes soft jade (like mutton fat jade) and hard jade (like emerald). In a broader sense, besides soft and hard jade, there are many colored stones that count as jade, such as serpentine, turquoise, malachite, agate, crystal, amber, ruby and emerald.

- 冰种翡翠喜鹊报春茶具（清）
  Ice Emerald Teaware with Magpie Heralding Spring Pattern (Qing Dynasty, 1616-1911)

- 白玉三羊执壶（清）
  White Jade Pot with Handle and Three-sheep Pattern (Qing Dynasty, 1616-1911)

饮茶风大盛，开始出现玉质茶具，如河南偃师杏园李归厚墓中的玉石杯。明代陵墓中出土的明神宗御用玉茶具，由玉碗、金碗盖和金托盘组成，玉碗底部有一圈足，玉材青白色，洁润透明，壁薄如纸，光素无纹，工艺精致。清代皇室也常用玉杯、玉盏作茶具。

China has a long history of jade craftworks. Tea was very popular in the Tang Dynasty, so jade teawares began to be in use, such as the jade cup from the tomb of Li Gui in Xingyuan Village, Yanshi, Henan Province. Jade teawares for royal families excavated from mausoleums of the Ming Dynasty include jade bowl, gold cover and base. The jade bowl has a round foot and the white green jade is clean, transparent, paper thin, and plain without any lines. This teaware is made with delicate skills. In the Qing Dynasty, royal members also used jade cups or bowls as tea wares.

- 金盖金托白玉杯（清）
White Jade Cup with Gold Cover and Base (Qing Dynasty, 1616-1911)

- 白玉菱花盏托、圆杯（清）
White Jade Cup with Water Caltrop and Saucer (Qing Dynasty, 1616-1911)

- 螭虎柄青玉碗（宋）
Bluish Jade Bowl with Loong and Tiger Handle (Song Dynasty, 960-1127)

老茶具的纹饰题材
The Themes of Patterns on Old Teawares

中国古代茶具的纹饰在不同时代有着不同的表现，多种多样的装饰题材反映出不同社会生活的审美意识。

几何纹，指的是以点、线、面组成的有规则的几何图形，是一种原始的装饰纹样，最早出现在新石器时代的陶器上，商周时期出现的原始瓷茶具上也有多种几何纹的装饰图案。而且在后世的茶具上，几何纹始终是一种常见的装饰图案或辅助纹饰。

Patterns on traditional teawares changed with times, which express different appreciation of social lives in different times.

Geometric patterns, refer to the regular patterns formed with points, lines or surfaces, are an original decoration pattern. They were firstly found on the ceramic utensils in the Neolithic Age. Ancient porcelains in the Shang and Zhou dynasties also had various geometric patterns. Throughout history, geometric patterns have been always common decoration or supplementary patterns on the teawares.

- 回纹彩陶钵（战国）
Colored Ceramic Bowl with Fret Pattern (Warring States Period, 476 B.C.-221 B.C.)

- 越窑云纹罂（五代）
Small Mouth Jar with Cloud Pattern from *Yue* Kiln (Five Dynasties, 907-960)

龙凤纹也是古代茶具上常见的纹饰。早期的龙纹为鳄鱼形状，神态凶猛，并没有神圣、威严的意味。从唐代开始，龙纹成为帝王的象征，龙的形象威武，极有神采，而且龙纹的使用也有了严格的限制。元代各种瓷茶具上都多见龙纹，一般龙头小，身为蛇形，有三爪、四爪、五爪之分，身形矫健灵活。明代龙纹身体粗壮，龙首有角，龙须上卷，身有鱼鳞。这时对使用龙纹的限制更加严格，五爪龙为皇帝专用，民窑很少有龙纹瓷器的制作。清代龙纹更加常见，顺治、康熙时的龙纹威严，带有盛世的气派。乾隆时开始，龙纹的限制逐渐放宽，龙的纹饰寓意大为世俗化。嘉庆以后的龙纹更无神圣之势，成为民间喜闻乐见的装饰题材。

The loong-and-phoenix pattern is also a common pattern on ancient teawares. Early loong patterns look like crocodiles, frightening and violent instead of sacred and dignified. Since the Tang Dynasty, the loong pattern became the symbol of emperors. Thus loong had a powerful and vigorous image and there were also restrictions about using its image. This pattern can be easily found on porcelain teawares of the Yuan Dynasty. Generally speaking, these loong patterns with small heads have three, four or five claws, which look robust and agile. Loong patterns in the Ming Dynasty have strong bodies, heads with horns, rolling-up whiskers and scales. At this time, restrictions on the usage of loong patterns were more rigid. Five-clawed loong was special for the emperor, so folk kilns seldom made porcelains with loong patterns. This pattern became more common in the Qing Dynasty. The loong patterns in the Shunzhi and Kangx periods were ever dignified, which manifested the flourishing age at that time. The Qianlong Period witnessed the relaxing of restrictions on the usage of loong patterns, so their meaning became more secular. After the reign of Emperor Jiaqing, loong patterns no longer expressed any scaring meaning and became popular decoration subjects among ordinary people.

- 龙首三足紫砂壶（清）
  Purple Clay Pot with Tripod and Loong Head Spout (Qing Dynasty, 1616-1911)

- 掐丝珐琅飞凤茶壶（明）
  Filigree Enamel Tea Pot with Flying Phoenix Pattern (Ming Dynasty, 1368-1644)

在古代茶具上，一些日常生活中常见的动物图案，在装饰题材中一直占有主要地位。例如具有吉祥寓意的鸟纹、鱼藻纹、蝴蝶纹等。

植物纹样在古代茶具上较为多见的有蕉叶纹、团花纹、卷草纹、缠枝纹、折枝花纹、瓜果纹等。从宋元时期起，每个朝代都有自己特有的植物纹样，明清时期植物纹样的运用非常广泛。

Animal patterns were always an important part in the decorations on ancient teawares, especially those with auspicious meanings, such as bird, fish swimming among algae and butterfly scenes.

The most common plant patterns on ancient teawares are banana leaves, flowers, rolling grass, winding branches, broken branches and melons. Since the Song and Yuan dynasties, every dynasty had its unique plant pattern, and in the Ming and Qing dynasties, plant patterns were more broadly used.

- 青花鱼藻纹杯（明）
Blue-and-white Cup with Fish and Algae Pattern (Ming Dynasty, 1368-1644)

- 青花花鸟纹茶壶（明）
Blue-and-white Tea Pot with Flower and Bird Pattern (Ming Dynasty, 1368-1644)

人物历来是瓷器上被描绘最多的一类图案，古代茶具上的人物图案，出现最多的要数婴戏纹，在明清时期的彩绘瓷茶具上，婴戏纹十分常见，而且内容丰富。孩童天真烂漫的形象折射出人们对社会安定、多子多福的祈盼。除婴戏图外，常见的人物纹样还有渔樵耕读、仕女图、八仙图、历史故事图等。

Personage is the most used pattern on porcelains. Among these personal character patterns, children at play is the most common one. Children playing scenes on painted porcelain in the Ming and Qing dynasties are the most numerous. The naive and happy image of children reflects people's wish for a stable society and a happy family. Other common personage patterns include fishers, lumberjacks, farmers and scholars, beautiful women, the eight immortals and scenes from historical stories.

- 青花"赤壁放游"纹提梁壶（清）
  Blue-and-white Loop Handled Pot with "Touring at the Red Cliff" Scene (Qing Dynasty, 1616-1911)

- 青花八仙人物碗（清）
  Blue-and-white Bowl with the Eight Immortals (Qing Dynasty, 1616-1911)

> 竹木茶具

　　竹木茶具是指用竹或木制成的茶具。用竹制成的茶具，大多为辅助用具，如竹夹、竹瓢、竹茶盒、竹茶筛、竹茶炉等；而用木制成的茶具，多用作盛器，如碗、具列、涤方等。竹木茶具轻便实用，取材容易，做工简单，多为民间百姓家使用，但其中精工细雕者，也出现

• 木制茶箱（清）
Wood Tea Case (Qing Dynasty, 1616-1911)

> Bamboo-wood Teawares

Bamboo-wood teawares refer to those made of bamboo and wood. Bamboo teawares are often supplementary ones, such as bamboo tongs, ladles, tea boxes, tea sieves and stoves. Wood teawares are often used as containers, such as bowls, shelves and rinsing boxes. Bamboo and wood teawares are handy and practical with common materials and easy processing. Therefore, they are mostly used by ordinary people. However, some delicately carved teawares could also be found in big officials' home and rich families.

　　The mid stage of the Tang Dynasty began to see bamboo and wood teawares. Lu Yu alone recorded a dozen of the bamboo and wood teawares in *The Classic of Tea*. Such teawares in the Song Dynasty mostly followed the features of Tang and developed a wood box to store tea. In the Ming and Qing

在达官贵人之家。

竹木茶具形成于中唐时期，陆羽在《茶经》中所载的竹木茶具就达十余种。宋代的竹木茶具大多沿袭了唐代的习惯，并发展出贮茶用的木盒。明清两代，饮用散茶之风盛行，竹木茶具种类减少，但工艺更为精湛。明代的竹茶炉、竹架、竹茶笼，以及清代的檀木锡胆贮茶盒等传世精品均为例证。近代和现代以来，在少数民族地区，竹木茶具仍占有一定的位置，如哈尼族和傣族的竹茶筒、竹茶杯；藏族和蒙古族的木茶碗、木茶槌；布朗族的鲜粗毛竹煮水茶筒等。

dynasties, loose tea became popular. Thus bamboo and wood tewares decreased in variety but developed better skills. Many masterpieces passed down, such as bamboo stoves, shelves, cages in the Ming Dynasty and the ebony tea box with tin base in the Qing Dynasty. In modern and contemporary times, bamboo and wood tewares still have an important place in many areas inhabited by ethnic groups, such as the bamboo tea tube and cup of the Hani and Dai ethnic groups; wooded tea bowl and hammer of the Tibetan and Mongolian ethnic groups; fresh coarse bamboo tea tube to boil water of Blang ethnic groups; etc.

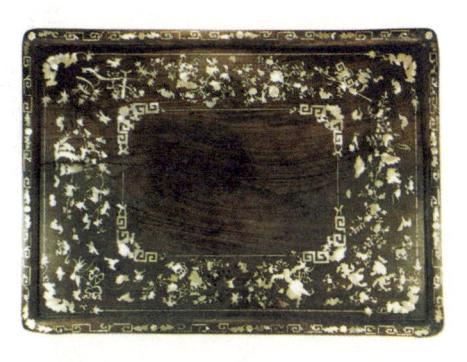

- 翻簧竹茶壶桶（清）

Bamboo Tea Pot Bucket with Reversing Spring (Qing Dynasty, 1616-1911)

- 木嵌螺钿茶盘（清）

嵌螺钿，是指用螺壳与海贝的珠光层磨制成人物、花鸟图形或文字等薄片，根据需要镶嵌在器物表面的装饰工艺。

Wood Tea Plate with Shell Inlaiding (Qing Dynasty 1616-1911)

Shell inlaiding means decorating the utensils with personage, flower, bird pattern or characters made of thin slices of seashells' inner surface.

少数民族的茶饮与茶具
Tea and Teawares of Ethnic Groups

藏族的酥油茶

酥油茶对藏族人来说如同饭食一样重要。酥油茶的做法是把茶砖切开捣碎，加适量的水煮沸后滤出茶渣，调入食用酥油，茶汁和酥油就混合成乳白色的"酥油茶"。打酥油茶用的茶桶多为铜制，也有用银制作的。而茶碗以木碗居多，但常常用金、银、铜等镶嵌装饰。还有用华丽而昂贵的翡翠制成的茶碗，常常被当作传家之宝。

Butter Tea of Tibetans

For Tibetans, butter tea is as important as their main food. The process of making butter tea is: cut and crush the tea bricks; add an appropriate amount of water and boil; filter the dregs and mix with butter. The milk white mixture of tea fluid and butter is "butter tea". Most tea buckets for making butter tea are made of copper, while some are made of silver. People often use wood bowls but decorate them with gold, silver and copper. Some tea bowls are even made of emerald and become heirlooms.

● 打酥油茶的藏族人
Tibetan Making Butter Tea

拉祜族的烤茶

拉祜族自古有饮用"烤茶"的习俗，将特制的小土陶罐在火上烤热后，放入茶叶，然后不断抖动小陶罐，使茶叶受热均匀，慢慢膨胀变黄，再将少许沸水冲入陶罐内，罐内顿时泡沫沸腾，茶香四溢。泡沫散去后，再加入大量开水烧煮片刻即可饮用。茶具以陶器、竹器为主，有烤茶罐、分茶壶、饮用杯等器具。

Baked Tea of Lahu People
Lahu people have a traditional habit of having "baked tea". They heat a special small ceramic jar on the fire and put in tea. Then they shake the jar to heat the tea leaves evenly. When the leaves swell and become yellow, people will add a small amount of boiling water. Boiling bubbles will come out at once with great tea fragrance. After the bubbles fade away, add a large amount of boiling water and it is ready to serve. Lahu people use ceramic and bamboo utensils, such as tea baking jar, tea dividing pot and drinking cup.

蒙古族的奶茶
蒙古族所喝的茶主要是奶茶，又名"蒙古茶"。蒙古族人每天早上第一件事就是煮奶茶。将水烧开后冲入放有茶末的净壶或锅，慢火煮2~3分钟，再将鲜奶和盐兑入，烧开即可。奶茶在饮用时还可加入黄油、奶皮子、炒米等，其味道芳香可口。烹制奶茶使用的茶具比较独特，主要有搅茶臼及木槌，搅茶臼呈倒圆锥形，木制，而木槌为搅茶用具。蒙古族早期是用木皮当碗，后来发展到大量用椴木碗，也有用桦木制成的。有些木碗用白银镶嵌装饰，外面刻有传统的花纹，是富人家常用的茶具，现在多用景德镇烧制的印有龙形花纹的细瓷碗。茶壶多为铜制，亦有银制的，造型别致。多数为圆形或椭圆形，嘴小，底大，外表锃亮，经常在壶盖、提手或壶嘴上镶嵌花纹图案。

- 蒙古族制作奶茶用的多穆壶
  Domo Pot for Mongolians' Milk Tea Making

Milk Tea of the Mongolians
The Mongolians' first thing in the morning is to make milk tea. They pour boiling water into a clean pot or cauldron with tea powder and simmer for two to three minutes. Then they will add fresh milk, salt, and cook till boiling. People may also add butter, cheese slices or fried millet, which are very delicious. Milk tea uses special teawares — tea stirring mortar and wood hammer. The tea stirring mortar is in a reversing cone shape, made of wood while the wood hammer is for stirring tea. In early times, the Mongolians used wood barks as bowls and later they used basswood and birch bowls. Some wood bowls were inlaid with white silver and engraved with

traditional patterns, which were for the rich families. Now, most Mongolian people prefer fine porcelain bowls with loong pattern made by *Jingdezhen* Kiln. Most of the tea pots are made of copper or sometimes silver, which have unique styles. Most of the tea pots are round or oval in shape, with a small spout, large bottom, bright outside surface. Craftsmen often carve patterns on the lid, handle or spout.

● 蒙古族的茶壶与茶杯
Mongolian Tea Pot and Cup

> 果壳茶具

果壳茶具是指用果壳制成的茶具，其工艺以雕琢为主。比如用手工将葫芦、椰子等硬质果壳加工成茶具，大多为水瓢、贮茶盒等辅助用具。葫芦水瓢多见于北方，而椰壳茶具主要产于海南岛。果壳茶具虽比较少见，但其历史十分悠久，

> Nutshell Teasets

Shell teawares refer to those made of nut shells, which stress on carving and polishing. For example, most handmade teawares use hard shells of gourd or coconut are used as supplementary tea wares such as ladles or tea boxes. Gourd ladles are more common in the northern part while coconut-shell teawares are

- 椰壳雕博古图碗（清）
Coconut Shell Carving Bowl with Ancient Auspicious Pattern (Qing Dynasty, 1616-1911)

- 锡胎椰壳雕茶杯（清）
Carved Coconut Shell Tea Cup with Tin Lining (Qing Dynasty, 1616-1911)

• 锡胎椰壳雕盖碗（清）
Carved Coconut Shell Tea Bowl with Tin Base (Qing Dynasty, 1616-1911)

• 锡胎椰壳雕茶壶（清）
Carved Coconut Shell Tea Pot with Tin Lining (Qing Dynasty, 1616-1911)

早在陆羽的《茶经》中就已描述过用葫芦制瓢的工艺，并历代沿用。椰壳茶具主要是作为工艺品，其外形黝黑，常雕刻山水或字画，内衬锡胆，能贮藏茶叶。

mainly produced on Hainan Island. Although shell tea wares are rarely seen, they have a long history. Lu Yu described the skill of making gourd ladle early in his work *The Classic of Tea* and this skill was passed down through dynasties. Coconut shell tea wares are always taken as artworks. They have black surfaces with carved mountain and river paintings or calligraphy. The inner part is made of tin and is good for storing tea leaves.

老茶具的收藏与保养
Storage and Maintenance of Ancient Teawares

　　流传至今的老茶具，记录了古人饮茶的情趣和韵味，受到爱茶人士和收藏家们的青睐，其中以瓷茶具和紫砂茶具的收藏最为普遍。对于这些历经岁月风尘的老茶具，正确而细致的养护十分重要。

Ancient teawares, which record the interest and charm of ancient people, are welcomed by tea lovers and collectors. Porcelain and purple clay teawares are the most popular ones. It is very important to carefully maintain these ancient teawares with the right methods.

> 瓷茶具的收藏与保养

储存瓷质老茶具最好的方法是放在定做的盒子里，盒子里衬上海绵或泡沫垫，尽量不要把两件瓷茶具放在一起。如果作为陈设，瓷

青花凤穿牡丹纹茶叶瓶（清）
Blue-and-white Tea Bottle with Phoenix Flying Between Peony Pattern (Qing Dynasty, 1616-1911)

> Storage and Maintenance of Porcelain Teawares

The best way to store old porcelain teawares is to put them in specially made boxes with a sponge or foam mat inside. It is better not to put two pieces in one box. If the teaware is for display, fix it on a wood shelf instead of a glass display stand. To avoid unexpected collision, precious porcelain teawares can also use nylon wire to fix its cover.

Try to use clean and dry hands while appreciating old porcelain teawares and take off accessories like rings, since they may scratch the glaze on the porcelain. Do not wear gloves while holding the porcelains because they can fall off easily from gloves. When taking a teaware with base or lid, try to take them separately. Do not hold them at the same time, so as to prevent them from falling or breaking. Use both hands to hold big

具最好放在固定的木架子上，比如实木做的博古架，而尽量不要用玻璃做的陈列架。为防止磕碰，在展示珍贵的瓷茶具时还可用透明的尼龙线固定上部。

在把玩瓷质老茶具的时候，双手应该保持洁净和干燥，取下戒指等饰品，因为坚硬物很容易划伤瓷器的釉面。拿瓷器时不要戴手套，因为这样瓷器很容易从手中滑落。在拿起一件带座、带盖的瓷茶具时，应将座、盖和主体分别单拿单放，不要连盖带座一起端，防止移动时脱落打碎。拿起器形较大的瓷茶具时，一定要双手捧握，不要单手拿盘、碗的一边，以防断裂。几个人同时鉴赏时，一个人欣赏完毕，把器物放到桌上，下一个人再拿起来欣赏，不要两人手递手地传看，以防在传递中失手脱落。

平时可以用湿布轻轻擦拭瓷器上的灰尘，或用柔软的刷子清扫瓷器，轻刷瓷器的缝隙。不要用水直接清洗未上釉的陶器，因为陶器有吸水性，而且出土的陶器外表很松垮，有的器身的化妆土已经呈粉末状，不能清洗。

porcelain tearwares. Do not hold one side of the plate or bowl, or they may break. When several people appreciate a piece together, one should put the ware on the table first and the next one should take it to continue. Do not pass wares from hand to hand or they may fall in between.

Clean the dust on the porcelain gently with a wet rag, or use a soft brush to clean the seams on the wares. Do not clean unglazed potteries directly with water. Since potteries absorb water and the surface of excavated potteries is very crisp, some engobe may have become powdered, so direct washing should be avoided.

• 磁州窑白地黑花茶叶瓶（明）
White Tea Jar with Black Flower Pattern from *Cizhou* Kiln (Ming Dynasty, 1368-1644)

> 紫砂茶具的收藏与保养

收藏紫砂茶具，日常使用中的保养非常重要。事实上，一把紫砂壶烧成后，由于胎骨沾染火气，紫砂间微孔结构松散，壶体很脆，容易受热胀冷缩的不良影响，而通过"养"壶，可以改善其结构特性。有些人出于珍视，把紫砂壶束之高阁，只藏而不养，实在是暴殄天物，其实养壶的目的在于增强紫砂壶"蕴味育香"的功能，使紫砂壶愈用愈光亮，能够尽显高雅品位。

既使再好的紫砂壶，不养也是没有光泽的。特别是有着书画陶刻装饰的旧壶、古壶，久养之后，纹样的立体感会得到加强，使古雅之气更加浓郁。

要使紫砂壶能够在使用过程中不致磨损，并能长期保持美观，就

> Storage and Maintenance of Purple Clay Teasets

Daily maintenance is crucial in storing purple clay tea sets. In fact, a finished purple clay pot is very fragile and the spaces inside the clay become loose since the pot is fired. Thus, it may expand with heat and contract with cold. The maintenance of tea pots can improve its structural features. Some people treasure it and store the pot on a high shelf. This is actually ruining it. In fact, the purpose of maintaining the pot is to upgrade its "fragrance" and "brightness" and to show its elegance.

Purple clay pot will not look bright if collectors do not maintain it, no matter how outstanding the pot is. Especially those ancient pots with carved paintings and calligraphy, they will have more stereoscopic patterns and elegant feeling

要有一套正确的养壶方法。最基本的养壶原则有以下几点：

1. 不管是新壶还是旧壶，使用时都要先把壶身上的各种污垢清除干净。旧的紫砂壶在泡茶前，应先用沸水冲烫一下。在启用一把新紫砂壶时，应先用水将紫砂壶里外洗干净，放入无油污的锅里，加入水煮沸。水沸后加入茶叶，不久就可以熄火，用余热焖壶，待茶水稍凉，捞尽茶叶，再点火煮沸，如此三四次，就可以使新壶的土味除去，也使新壶初次受到茶叶的滋润。此时将紫砂壶取出晾干，就可以沏茶使用了。

2. 紫砂壶最怕油污，用上油的方法来产生光泽是不可取的。平时

after long time maintenance.

A proper method to maintain purple clay pots is necessary to keep it from wear and tear and to hold its beautiful look. Here are some basic principles in pot maintenance:

1. Both new and old pots should be clear from dirt before use. Use boiling water to wash old purple clay pot before making tea. When use a new pot, clean the inside and outside with water. Put the pot into a cauldron without grease and boil it with water. Add tea after boiling and extinguish the fire soon after. Use this temperature to simmer the pot. Take the tea out when the water is cooler and boil it again. Repeat this process three or four times. Then the clay smell will go away and it can moisten the new pot with

- 黄玉麟制紫砂方斗壶（清）
Square Purple Clay Pot by Huang Yulin (Qing Dynasty, 1616-1911)

- 紫砂龙头八卦一捆竹壶（清）
Purple Clay Pot with Bamboo-shaped Wall, Eight Diagrams Pattern and Loong Spout (Qing Dynasty, 1616-1911)

使用或保存时也要远离有油的物品。如果沾上油污，必须马上清洗，否则就会导致土胎留下油痕，影响其对茶汤的吸收。清洗方法是用细布蘸肥皂轻轻擦拭，然后用手摩挲，让壶体现出本色。

3. 泡茶次数越多，壶吸收的茶汁就越多，吸收到一定程度，就会渗透到壶表，使之发出润泽如玉的光亮来。沏茶时，紫砂壶的表面温度很高，可以用湿布擦拭壶体，反复多次，直至壶温降下来。还可以将头道茶水倒入杯中，往壶中加入沸水，然后把头道茶水倾倒在壶体上，以茶汤养壶，待壶不烫手时改

tea at the same time.

2. Greasy dirt is the biggest danger to purple clay pots, so do not try applying oil to make the pot look brighter. Keep the pot away from oil when use and maintain it. Greasy dirt should be cleaned at once, or it will leave marks and affect its absorption of tea infusion. Use fine cloth with soap to wipe it gently and lightly stroke it with hands to manifest its original look.

3. The more the pot is used, the more tea it will absorb. When the pot absorbs a certain amount of tea, the tea will permeate to the surface and give it a jade-like brightness. When making tea, the surface temperature of the purple clay pot is very high. Wipe it with wet cloth for many times until it cools down. Or pour

• 百果壶（清）
Tea Pot with Fruit Decoration (Qing Dynasty, 1616-1911)

• 荷莲寿字壶（清 乾隆）
Tea Pot with Lotus Pattern and Carved *Shou* (Longevity) (Qianlong Period of the Qing Dynasty)

用手摩挲。另外，还可以将泡茶时留在壶中的茶渣取出，在壶体周身润擦，既可擦去壶身上的茶垢痕，也易使壶体光润亮泽。

4. 紫砂壶需要擦洗时，可用软毛小刷子或干净的棉布擦拭，然后用开水冲净，最后用清洁的茶巾稍加擦拭就可以了。

5. 有人在紫砂壶用过后，常常将茶渣或剩余茶汤留存在壶内，认为这样可以养壶。其实这样时间一久，壶内就会繁衍细菌，产生恶臭和酸馊味，对人对壶都有害。紫砂壶使用后应及时清理，将残茶倒出，用热水冲洗干净壶身内外，

the first round of tea in the cup and add boiling water into the pot. Then pour the water in the cup to the body of the pot in order to maintain the pot with tea. Stroke it gently with hands when the pot cools down. In addition, wipe the pot with the tea dreg in the pot to both clean the dirt on the pot and brighten its body.

4. Use a small soft brush or some clean cotton to wipe the pot when needed. Then wash it with boiling water and wipe the pot dry with a clean tea towel.

5. Some people leave tea dreg or infusion in the pot as they reckon this is good for the pot. Actually, after a period of time, bacteria will grow in the pot and emit odor and sour smell,

- 紫砂四方倭角隐竹顶壶（20世纪上半叶）
Purple Clay Square Tea Pot with Bamboo Design (First Half of the 20th Century)

- 紫砂提梁壶（清 雍正）
Purple Clay Loop Handle Tea Pot (Yongzheng Period of the Qing Dynasty)

不要在壶身留下水渍或茶垢。另外还可以稍微打开壶盖，以便让壶内残余的水气尽快阴干，避免产生异味。若因一时疏忽，使壶内的剩茶发生霉变，产生异味，可以在清除霉变的茶叶渣之后，往壶中注满开水，焖一会儿，把开水倒出，然后将紫砂壶浸在凉水中，反复两至三次，就会消除异味。

6. 使用一段时间后，紫砂壶也需要休息一下。将紫砂壶清洗干净后，先放在通风之处晾干，使土胎彻底干燥。存放时，最好能够密闭储藏。要尽量避免将壶放在灰尘、油烟过多的地方，以免壶面的润泽感受到影响。

which is harmful for both people and the pot. Clean the pot as soon as it is used by taking the dreg out and cleaning the inside and outside with hot water. Do not leave water stain or dreg behind. In addition, open the lid a little bit in order to dry the pot quickly as well as prevent it from having odor. If the pot is mildewed and smelly due to carelessness, fill the pot with boiling water after taking out tea dreg. Keep the boiling water for a while, pour it out and soak the pot in cold water. Repeat this process two or three times, then the odor will go away.

6. Give your purple clay tea pot a rest after using it for a period of time. Dry out the clean tea pot in a ventilated place. Try to seal the pot when storing. Try to avoid dust or grease in storing the pot, or the bright glaze may be damaged.

> 金属茶具的收藏与保养

金属茶具在收藏和使用过程中应该注意以下几个事项：

第一，在使用金属茶具冲泡时，应该注意壶内外的温度差，采用先暖壶、再冲泡的方法，使金属茶具受热均匀。第二，使用完金属

> Storage and Maintenance of Metal Tewares

Here are some tips in the storage and maintenance of metal teawares:

Firstly, pay attention to the difference of temperature between the inside and outside of the metal pot. Warm the pot before making tea, so the pot will absorb heat evenly. Secondly, clean

- 镶绿松石金执壶
Gold Pot with Handle and Inlaid Green Tophus

- 铜荷花瓣托盏（宋）
Copper Bowl and Base with Lotus Leaf Pattern (Song Dynasty, 960-1279)

茶具之后，应该将茶具清洗干净，不要留下茶渍，以免腐蚀金属。第三，在清洁金属茶具时，用柔软的干布轻抹，尽量不要用化学类洗涤剂清洁，因为这些化学物质会与金属产生一定的化学作用，可能产生对人体有害的物质。第四，存放和收藏金属茶具时，要将茶具擦干或烘干，不要存放在有腐蚀性气体和潮湿的地方，避免金属生锈。还要注意金属茶具不要与其他硬物一同存放，以防其碰撞茶具，损坏表面。

the teawares after using. Do not leave dreg inside to avoid corrosion. Thirdly, try to use some soft dry cloth instead of chemical cleaners to clean metal teawares because these chemical substances may react with the metal and produce harmful substances. Finally, dry the teawares before storage. Do not keep them in humid places with corrosive gases to avoid rusting. In addition, do not keep metal teawares together with other hard items or they may collide and damage the surface.

> 漆茶具的收藏与保养

收藏漆器茶具应注意对其进行有效的养护,才能使其保持风采。第一,漆器不适宜温度和湿度的急剧变化,避免忽干忽湿,漆器茶具适宜放在温度、湿度恒定的环境

> Storage and Maintenance of Lacquer Teawares

To keep the splendure of lacquer teawares, effective maintenance is crucial. Firstly, do not keep them in places with rapid change of temperature, humidity, or make it dry or wet in a sudden. Lacquer teawares should be kept in places of

- 黑漆描金花卉纹茶壶桶(清)
Black Glazed Tea Pot Bucket with Gold Flower Patterns (Qing Dynasty, 1616-1911)

- 紫砂黑漆描金彩绘方壶(清 雍正)
Purple Clay Square Tea Pot with Colored Painting and Gold Drawing on Black Lacquer (Yongzheng Period of the Qing Dynasty)

内。第二，存放漆器茶具时，不要将其放得离地面太近，以避免吸收地面湿气，否则会使茶具脱漆、发霉。同时，避免阳光暴晒，这会使漆器出现变形、断裂。第三，在使用漆器茶具时，应该注意要轻拿轻放，不要与坚硬、锐利的物体碰撞或摩擦，避免剧烈的震动。第四，注意防尘。如果漆器茶具的表面有灰尘沉积，可以用棉纱布擦拭，或用柔软的毛刷清理，如果沾有污垢，可以用棉纱布蘸上少许食用油轻轻擦拭。

constant temperature and humidity. Secondly, keep lacquer tea wares away from ground or it may absorb moisture, depaint and mildew. At the same time, keep the teawares from blazing sunshine or it can lead to a deformation or break of the wares. Thirdly, take and set them gently. Do not collide or chafe them with hard or sharp items or shake violently. Last but not least, keep the teawares away from dust. Clean the dusty surface with gauze or a soft brush. If there are stains, wipe it lightly with gauze and cooking oil.

> 竹木茶具的收藏与保养

竹木茶具在使用过程中应尽量避免冷热温度的骤变，特别是竹质的茶具，受到骤冷骤热的刺激很容易产生

> Storage and Maintenance of Bamboo - Wood Teawares

Try to avoid a sudden change of temperature when using bamboo and wood teawares, for they may crack due to

- 瘿木手提式茶簏（清 乾隆）
Burl Wood Tea Set Basket (Qianlong Period of the Qing Dynasty)

- 竹木茶具
Bamboo-wood Tea Set

爆裂现象。其次，在存放的过程中尽量避免过于干燥，因为竹木的材质长期干燥会变形乃至开裂。再有就是尽量避免阳光的直晒，这样也会造成器物的变形损坏。

a rapid change in temperature, especially bamboo ones. Do not store them in a very dry place for a long time because bamboo and wood can deform or break. Also, try to avoid direct sunshine.